BEST of Mathematics in School

Selected articles from *Mathematics in School*

Edited by

Michael Cornelius

Longman

Published by Longman Group UK Limited, 6th Floor, Westgate House, The High, Harlow, Essex CM20 1YR, England.

British Library Cataloguing in Publication Data

Best of "mathematics in school"
1. Great Britain. Schools. Curriculum subjects: Mathematics
I. Cornelius, Michael
510′.7′1041

ISBN 0-582-04088-4

ISBN 0-582-04088-4

Printed in Great Britain by Bell and Bain Ltd., Glasgow

Contents

1 Introduction
by Michael Cornelius

Mathematics in School first appeared in November, 1971 with the stated main aim 'to provide teachers of 7–16 year old children with practical and interesting material which has a direct connection with their work in the classroom'. The first editorial explained '. . .we hope to print work in progress which will benefit by public debate. . . we offer a platform where ideas can be floated and where prejudices can be aired'.

The journal is now well established and during the past two decades it has contained a wealth of discussion and ideas on the teaching of mathematics with contributions from a host of individuals ranging from school pupils and student teachers to university professors. Within the editions which have been published there are numerous articles which are likely to be of interest to the teacher of the 1990s and this volume attempts to draw attention to material which may be of use to teachers in today's classrooms — many such teachers will probably be unaware of much that has appeared in *Mathematics in School* in past years.

Throughout its period of publication the journal has been fortunate in having three dedicated and gifted editors, all of these: Peter Reynolds, 'Bud' Winteridge and David Neal have selected material for sections of this book. The other sections represent the choices of the current editorial team and thus we have a collection of articles chosen by six people all of whom have had long and close connections with *Mathematics in School*.

The selection of material to be included was not easy and there were many articles which the contributors would have liked to include had more space been available. Final choices were difficult but it is hoped that readers will find things here which are of interest and, above all, useful to the classroom teacher. With the impending introduction of the National Curriculum in schools in England and Wales and in the light of the recent (1988) report of the Mathematics Working Group, it is illuminating to discover that many of the 'current' issues were considered and discussed in the columns of *Mathematics in School* in past years. Modern technology has made an impact on the teaching of mathematics but the basic issues and problems facing the *classroom* teacher have remained the same.

Mathematics in School is now an established, respected journal. The following chapters give a sample of some of the content over nearly twenty years — if readers enjoy them and subsequently find themselves spurred on to reread and consult old editions then the contributors to this volume will feel that they have achieved a useful goal.

2 Primary Mathematics
by Lesley Jones

Mathematics in School aims to provide material for teachers of children aged 7–16. In most issues the focus of interest is clearly towards the upper end of this age group. There are some items which are directly applicable to the primary classroom but they are often few and far between. Many articles deal with mathematical topics which are themselves of interest to teachers — it was beyond the scope of this section to select such items. What I have tried to do is to select a variety of articles which will have some relevance to the primary teacher. My approach was along the lines of *Desert Island Discs*. I tried to imagine a primary teacher setting out on his/her profession and make a selection which might stimulate and support them in their early years of teaching.

Ernest Choat provides a scholarly discussion of Skemp's work on understanding; he relates his account closely to the primary classroom and discusses the implications of Skemp's work for the primary teacher. Gill and Gill discuss the importance of examining attitudes towards ethnic minorities in this country and demonstrate ways in which mathematical examples and topics can be chosen which show that we value the knowledge gained in a variety of cultural settings. We are reminded by Tom Brissenden of the importance of discussion in mathematics and he provides some ideas about suitable tasks which will provoke and stimulate a discussion about the process of mathematics.

Donald Thornley suggests some stimulating ideas for investigations in the primary classroom whilst the charming story of Tanya Gram illustrates the way in which a mathematical problem solving situation can so easily spread across the curriculum.

Finally Anita Straker reminds us of the need to practise *mental mathematics* and proposes some alternative ways of achieving this.

Understanding in Young Children's Mathematics

by Ernest Choat, Goldsmiths' College, University of London

Occasionally, a paper that has a profound effect on the theory of education is published. The elaborated and restricted codes in language postulated by Bernstein is an example. Although teachers may be aware intuitively of the circumstances (even Piaget's deliberations on children's stages of thought development is in this category), someone has to specifically detail the facts before significance is realised. Skemp[1] has done this with "understanding in mathematics".

Skemp initially refers to two forms of understanding in mathematics — relational understanding and instrumental understanding. Relational understanding is not only knowing what to do but why as it includes rationalisation of the underlying mathematical relationships and properties. Instrumental understanding is when pupils and teachers think that they understand something if they are able to obtain correct answers to a given category of questions without knowing why the method works. The connotations attached to these definitions have a significant bearing on the way children's mathematical competence is assessed (Choat[2]). Are children assessed in mathematics for knowing what they are doing, or are they assessed for being able to do without knowing? The level at which children understand must reflect the way they have learned. Moreover, understanding should not be confused with skill. Skill is the capability to perform a set routine, and requires a period of experiment and practice by trial and error before the routine is carried out smoothly and efficiently. A skill may be performed with either relational or instrumental understanding. Furthermore, although a teacher may believe that she is teaching mathematics, relational mathematics and instrumental mathematics are effectively two different subjects; the former requiring children to apply thought to rationalise, the latter being automatic processes.

Skemp[1] suggests that many teachers may teach instrumental mathematics because it has certain advantages. Instrumental mathematics, within its own context, is easier to grasp than relational mathematics. Rules are given for compliance and are remembered easily, and a page of correct answers is obtained more quickly. Consequently, rewards are immediate and apparent for the satisfaction that this gives to children cannot be under-rated.

Although less knowledge is involved with instrumental mathematics for a child to secure the right answer more quickly and reliably, relational mathematics also has advantages. It is more adaptable to new tasks by knowing not only which method works but why it works, and enables a child to adapt methods to new problems. Instrumental understanding necessitates memorising which methods apply to certain problems, and also learning a different method for a new problem. This practice with children is rife when the 11+ examination prevails. The mathematics questions follow the same format year after year, and teachers teach the knack of how to answer each question. On the other hand, according to Skemp, although it may seem a paradox relational understanding is certainly harder to learn but easier to remember. This arises because there is more to learn with relational understanding — connected wholes as well as rules — but once learned the result is more lasting owing to less re-learning, and the time taken may be less in the long term. Relational knowledge can be effective as a goal also as the need for external rewards and punishments is reduced to make the motivational aspects of a teacher's job easier. If children receive satisfaction from relational understanding, they may not only try to understand new material relationally, but seek new material and new areas on their own initiative. Skemp considers therefore that a case might exist for instrumental mathematics in the short-term, and within a limited context in the long term, but not in the context of a child's whole education.

Skemp's theory has been subjected to comment and amendment both by himself and others, namely Skemp[3,4], Byers and Herscovics[5], Backhouse[6], Buxton[7], and Choat[2], but preliminary explanations must precede digression of these considerations. Learning is a measurable change that involves either the acquisition of some new body of information (cognitive learning) or a shift in values, attitudes, interests, or motivation (non-cognitive learning) (Choat[2] p. 92). Consequently, learning may cover many different activities that, although they have the common characteristic of a change in behaviour, are radically different and, according to Thouless[8], these are the acquisition of information by mechanical remembering, the acquisition of information by logical remembering, the acquisition of certain bodily skills, the acquisition of a number of intellectual skills, and the acquisition of a certain group of desirable or obligatory attitudes that include aesthetic, social, and moral attitudes. Learning therefore takes place through the interaction between the learner and some other object, person, or experience, and for young children this is facilitated by play and the environment.

Play is necessary to cater for children's social, emotional, and physical development but it is also an important means for providing intellectual development. As Isaacs[9] relates, play provides the normal growth in manipulative skills, and imaginative expression in discovery, thought, and reasoning. The teacher's task is to recognise the spontaneous interests and activities of children at each successive age and provide the opportunities, materials and stimuli that will allow the greatest fulfilment along all directions of their growth. Geometry is the basis of much of children's play through shape, size, position, balance, and movement when it provides them with opportunities to experiment, manipulate, and control their own bodies as well as discovering the capabilities and properties of objects. It could be said therefore that mathematics originates a great deal of children's play so it stands to reason that through play children are learning mathematics. The situation could be likened to the chicken and egg as both are dependent on the other for their existence, but a child could play and not learn mathematics.

The environment, too, as Choat[2] explains, is vital to children's mathematical development as they live in a series of situations that it creates, and the interaction with these situations constitutes experience. During this interaction of activities and explorations, concepts are formed. Consequently, children's educators (teachers and parents) are responsible for determining the environment and, subsequently, the experiences which children acquire. Nevertheless, concept formation is not an automatic process. Gathering and piecing together experiences are possible through perception, but perception is not a casual survey of the environment. It is diagnosing the objects that constitute the environment through touch, sight, sound, taste, smell, balance, and movement; all of which may be affected by the condition of the learner (needs, interests, growth, attitude, emotional state, desire at the moment of perception, strength of the stimuli, and likelihood of occurrence). Perceptual space is geometric, so early mathematical knowledge originates from contact with objects which constitute the environment, and play is a constituent of the environment. Perception enables children to comprehend spatial relationships, understand objects' movements, and co-ordinate their own movements with those of the objects. Children who use their perceptual powers act among things and not only on things to discriminate and conceptualise by image and thought.

Thus, the patterns of relationship that reveal mathematics to young children are acquired through experiences derived from the environment, with play being a major factor within environment, and the path towards relational understanding can be traced. Children engage in any new activity equipped with previous knowledge which they direct to a situation, but learning does not materialise unless they apply perception to the circumstances that the environment imposes. As a result of the experience, new concepts are formed, but such experiences are physical experiences only. A stage is reached when physical experiences reach their limit and mental application is necessary if further learning is to occur. This implies the need to employ logic and mathematics independent of the physical objects. Logic enables the implementation of deductive reasoning, and mathematics the rationalisation of the reasoning. Thus, logico-mathematical experience is actions of the mind, and these actions are internalised symbolic operations that enable higher order concepts to be formed. As with concepts, logical thinking cannot be taught but is developed by the extent that experience requires children to apply their mental powers.

Relational understanding therefore provides for the formation of concepts in two categories — physical experience and logico-mathematical experience. Skemp[3,4] extends his original model to include these definitions terming the directing of physical actions as delta-one, and goal-directed mental activity as delta-two, but introduces a new mode of understanding — logical understanding. This is derived from Byers' and Herscovics' suggestion of the existence of formal understanding which they define as the ability to connect symbols and notation with relevant mathematical ideas, and combine these ideas into chains of logical reasoning. Skemp supports the hypothesis by quoting the example of Byers and Herscovics of a pupil's writing:

$$\begin{aligned} x+3 &= 7 \\ &= 7-3 \\ &= 4 \end{aligned}$$

The contention is that if a child can explain that "the top line says adding 3 to x gives 7 so to find x we take the 3 away" then he has relational understanding. If the exact *meaning* is pointed out and the child re-writes it correctly, this would be evidence that he is capable of logical understanding brought about by reflection. The chain of logical reasoning is the connection of the logical sequences of the operation, according to Byers and Herscovics, but is this so? To relationally understand the example quoted, a child must have acquired the logic of the operation of addition and be able to appreciate inclusion, i.e. the relationship of part to whole. These are mental acts that precede the manipulation of symbols and notation in written form, and this contention agrees with Backhouse that "form is a means of representing a concept by agreeing with certain rules", while Buxton maintains that the formal construction of proof can be learned by rote. It is fair to say that the rules by which the proof is established have been logically induced in the form, but these rules have been previously deduced in the mind and the logical sequence of recording adopted for convenience to diagnose children's capability at various stages of progression in that notion of mathematics. The fact that a child can write two independent number symbols in the prescribed form and produce a further symbol that denotes their additive value,

$$\text{i.e. } 2+3=5 \text{ or } \begin{array}{r} 2 \\ +3 \\ \hline 5 \end{array}$$

is no indication that he has acquired relational understanding of addition. Number is not something impressed on the mind by external forces, or the perception of objects, but is a product of the minds actions in the valuation of quantity. As such, number is a system of internally symbolised representations. To operate on number, therefore, implies mental operations on the symbolised forms but, before the ultimate level of realisation is reached, a child must proceed through many levels of relational understanding in which he employs logic to rationalise and deduce (Choat[10]).

Logic is implicit throughout relational understanding and does not apply only to inductions to regulate form, so agreement cannot be reached to accept a separate category of logical understanding as defined by Skemp.

Further divergence arises through a lack of distinction between meaning and understanding and adds to the rejection of formal understanding. Skemp uses the expression "if the exact *meaning* is pointed out". This implies the use of words to describe a situation but, as Choat[2] emphasises, meaning is not understanding. A child may be told the meaning of something by his teacher and be able to repeat comprehensibly what the teacher has said, but this does not imply that he understands the process relationally. He may be able to perform the operation that has been explained, i.e. complete the operation of subtraction by "borrowing", but he does not understand relationally why he is "borrowing". Meaningful learning is none other than a sophisticated form of rote learning when a teacher, by explaining the principles involved, believes a child has understood the process relationally because he can produce the required formal evidence. In fact, the learning has been instrumental and raises the question of whether meaningful rote learning is possible as opposed to rote learning. Rote learning is unconnected learning and memorising facts to be reproduced on demand. These facts are used only in situations in which they were learned. Meaningful rote learning occurs when, although devoid of relational understanding, realisation of facts emerge as the result of continual rote practice. The way some children learn the tables of multiplication is an example. The children are practised in the tables constantly until they realise what the tables mean. When situations that require use of the tables arise, i.e. area, the cost of a number of items, etc., the children have sufficient meaning to know that a correct answer can be provided by using the tables, but understanding how the answer is reached remains a mystery. A stage is reached when the rules associated with the tables are at their limit and children cannot progress further, but it could be contended that meaningful rote learning places children in a position to be able to understand at a later date. This may be so but children will need to revert to rationalisation of lower order concepts before forming higher order concepts, and this requires application of relational understanding. Rationalisation enables the logical basis of relationships between concepts to be reasoned. A schema must accommodate to a new situation, and if it fails the new situation cannot be interpreted (Skemp[12]). Rationalisation therefore is the necessary interpretive process to provide accommodation.

Byers and Herscovics also introduce intuitive understanding into the discussion. Intuitive understanding is referred to as the ability to solve a problem without prior analysis of the problem. Dantzig[11] describes intuition in mathematics as a compelling insight that enables mathematical progression without the need for logical deduction. Intuition is a "sixth sense" that provides an answer to a problem and as such does not provide relational understanding as a child cannot elucidate how he reached his conclusions. Neither is conceptualisation involved as the process cannot be traced back to a starting point or assimilated into a schema. Thus, intuition is devoid of background knowledge from which future knowledge originates, and in many instances is attributed to children when instead they had an inspired guess. Intuition therefore cannot be taught as Byers and Herscovics recommend as it either returns to relational understanding for the conceptual process or is fitted to rules to provide instrumental understanding.

These deliberations reveal an aspect that hitherto has not featured in the exchanges on the theory of understanding in mathematics. Relational understanding depends on rationalisation of mathematical relationships and properties through experience, both physical and logico-mathematical, for its acquisition, and necessitates children to be active, involved and challenged. Instrumental understanding is devoid of experience and relies on rules which are administered by instruction through language which may be either spoken by a teacher or read from textbooks, workbooks, or work cards. This distinction has important repercussions for the learning and teaching of mathematics to young children.

Nursery schools are brimming with activities which are usually classified as play but, as the previous comments suggest, contain a great deal of mathematics. In fact, more mathematics is encountered in nursery school than in any other educational establishment with the children busily engaged in activities described as creative play, imitative play, sand play, water play, model making, carpentry, painting, picture and pattern making in two and three dimensions, collage, block building, clay, dough, plasticine, the Home Corner and Wendy House, outdoor apparatus, cooking, shopping, weighing, measuring, sewing, bead-threading, paper cutting and folding, books, stories, and jingles. This is by no means a final list for many other examples could be added (Choat[13]). Consequently, although mainly restricted to physical experience owing to the mental development of the children, a policy of relational understanding is explicit but the provision of activities alone is insufficient in young children's acquisition of mathematics. Although they may have freedom to follow their own direction in play, children need help to translate what they are doing, and this translation comes about by interpretation through language from the teacher. Words are the first mathematical symbols encountered and learned by children. Words interpret and communicate the first conceptualised spatial relationships. In number, although the "oneness of one", "twoness of two", etc., may be conceptualised, the information must be communicated by symbols (words), e.g. "one", "two", etc., before relational understanding can be ascertained. Children's vocabulary development therefore should proceed alongside physical and logico-mathematical experiences. Initially names may not be of objects but the noises associated with them, e.g. puff-puff, tick-tick, quack-quack, etc., until a level of understanding is reached to discriminate between objects. In many instances, when unaware of the conventional name, young children will invent names themselves. This implies that whenever possible the teacher should be an active participant in young children's play and activities by acting as interpreter and guide when the play and activities are in progress. This calls upon the professionalism of a teacher to use her skills of observation and knowledge of child development for when to to take an active role, when to stand back, when to interact, when to advise, and when to interpret. A teacher, thereby, becomes an *associate* in children's play, activities and experiences. Her participation does not inhibit but enriches, and specification of an active role is contrary to intervention. However it is doubtful whether many nursery school teachers appreciate the role of mathematics in the activities of young children while, apart from Matthews and Matthews[14], most researchers ignore its existence also. Every pre-school programme has as one of its priorities the fostering of language (Sylva, Roy, and Painter[15]), but relational understanding in mathematics is equally important if children are to develop the capability to think.

Only a small proportion of children are fortunate enough to attend nursery school so the task of adjusting the majority of children to school rests with the reception class teacher in infants' school. Nevertheless, every child who enters infants' school has had some experience of mathematics as it has featured largely in his early life, but many children need help to adjust to an environment that contrasts with home. Apart from matters concerned with learning, some children have emotional and social problems (Choat[13] pp. 108-120), and the responsibility for overcoming these falls on the reception class teacher. In fact, the way that children settle-in during this period is vital because they are forming their attitude towards school and developing their outlook towards learning and, as far as mathematics is concerned, being influenced in their attitude towards mathematics. In some schools an instrumental approach is followed and mathematics treated as if it does not belong to the lives of children. They are expected to learn aspects of mathematics but are unable to relate why the learning is necessary. This is particularly prevalent when number is treated as an automatic process, or when children are expected to complete pages of a workbook without having experienced the preceding practical activities. This is in contrast to a reception class teacher who makes it her task to discover what mathematics each child has acquired and then provides opportunities for further development by relational methods. This means being aware of the mathematics that could arise from various activities and experiences, and of the psychological factors associated with mathematical learning. Although the teacher should have a basic framework of mathematics in her own mind, she must not force stipulated mathematical notions on children but consider how appropriate the mathematics in the activities and experiences may be to each child's level of development. During the reception year, and subsequent years in infants' school, children gradually become less dependent on objects and begin to abstract, but they must be allowed opportunities to work through the stages of mathematical progression before being able to apply logico-mathematical experience.

Thus, the initial enthusiasm of young children and their

subsequent attitude to mathematics can be affected drastically by the manner in which mathematics is viewed by their teacher. This can be either as a thought mechanism through relational understanding when children are encouraged to enquire and come to terms with their environment, or a more formal encounter through instrumental understanding with a subject known as mathematics. Inevitably, the formal approach to learning mathematics leads to formal teaching. Children are no longer considered to require practical activities to derive experiences and relational understanding and, instead, language is used as the medium of teaching, but this detachment of language from experience and rejection of activity, are causal factors for children not learning. Patterns of relationship that reveal mathematics are acquired through experience, and this implies that language by itself is less important to young children than their experiences.

During *association*, language is allied to experience but with non-association language is adopted as the sole teaching and learning medium. Language is the most common form of interaction between human beings and when a child has a reasonable command of language it is natural to communicate and respond with him through language. Owing to the closeness that this gives, the assumption is made that the child comprehends the verbal exchanges. This pre-supposes that the child can reason explanations and instructions in the abstract whereas he may need physical experience to rationalise a situation and may not have understood the language, while his thought may not be at the level to interpret what the language defines. Teachers may think that they can transplant their thoughts and ideas into the minds of children through language, but the response is negative if the children are unable to rationalise at the required level.

If instrumental methods through language are used exclusively as the means for children to learn mathematics, children are required to be in a constant state of abstraction. This is beyond those children who have not reached the level to which the abstractions relate so the children begin to despair and reject mathematics. Increasingly, the children have to struggle to keep up and, without the lower order concepts to derive relational understanding, they soon develop an antipathy towards mathematics. Teaching by language domination gains momentum when children reach junior school. The children are considered too old to need play in learning, or to need physical experience. A teacher is prone to set a task and, because she has told the children what to do, expect the task to be fulfilled. This may be instanced with scale drawing and modelling. The process of reduction is explained and the children challenged to make a model of the school. Measurements are made, but confusion arises once application of scale is necessary as the children have not had prior experiences of depiction. Alternatively, a teacher may refer a child to the explanation of a process in a textbook, or require him to follow the instructions on a work card. Obviously, these should be at the child's reading level, but, even when they are, he becomes confused if without relational understanding. Over-reliance on instrumental methods in junior schools does not encourage children to pursue mathematical consciousness. Conditioned to instrumental understanding, they come to regard mathematics as a mechanical process devoid of interest or enjoyment. Such is the transition from what they were doing when younger that for some children mathematics changes completely and becomes unrecognisable from what they thought it was previously.

It may be said, therefore, that children could encounter two completely different types of mathematics. The first is mathematics acquired through relational understanding as a means for developing thought. In this capacity, mathematics enables a child to build mental models by providing him with a structure of logical sequences for his actions. Mathematics thereby is an inherent component within his actions and the basis on which his thought derives. The second is mathematics taught formally through instrumental understanding that abides by rules often explained by others. This distinction has a bearing on children's competence in mathematics and may answer a concern expressed by the Assessment of Performance Unit[16]. The APU consider that most 11 year olds can do mathematics involving the more fundamental skills and concepts to which they have been introduced, but there is a sharp decline in performance when application of concepts is probed deeply, and when basic knowledge has to be applied in more complex settings or unfamiliar contexts. Does this imply a neglect of relational understanding at the expense of an over-emphasis of instrumental understanding?

References

1. Skemp, R. R. (1976) "Relational Understanding and Instrumental Understanding", *Mathematics Teaching*, **77**, 20-26.
2. Choat, E. (1980) *Mathematics and the Primary School Curriculum*, pp. 36-45, Windsor, NFER Pub. Co.
3. Skemp, R. R. (1979), "Goals of Learning and Qualities of Understanding", *Mathematics Teaching*, **88**, 44-49.
4. Skemp, R. R. (1979) *Intelligence, Learning, and Action*, Chichester, John Wiley.
5. Byers, V. and Herscovics, N. (1977) "Understanding School Mathematics", *Mathematics Teaching*, **81**, 24-27.
6. Backhouse, J. K. (1978) "Understanding School Mathematics — A Comment", *Mathematics Teaching*, **82**, 39-41.
7. Buxton, L. (1978) "Four Levels of Understanding", *Mathematics in School*, **7**, 4, 36.
8. Thouless, R. H. (1969) *Map of Educational Research*, Windsor, NFER Pub. Co.
9. Isaacs, S. (1937) *The Physical Aspects of Child Development*, p. 40, London, Evans.
10. Choat, E. (1979) "Operations on Number", *Mathematics in School*, **8**, 5, 2-4.
11. Dantzig, T. (1930) *Number: The Language of Science*, pp. 247-248, London, Allen and Unwin.
12. Skemp, R. R. (1971) *The Psychology of Learning Mathematics*, p. 44, Harmondsworth, Pelican.
13. Choat, E. (1978) *Children's Acquisition of Mathematics*, Windsor, NFER Pub. Co.
14. Matthews, G. and Matthews, J. (1978) *Early Mathematical Experiences Guides*, London, Addison-Wesley.
15. Sylva, K., Roy, C., and Painter, M. (1980) *Childwatching at Playgroup and Nursery School*, p. 77, London, Grant McIntyre.
16. Assessment of Performance Unit (1980), *Mathematical Development: Primary Survey Report No. 1*, London, HMSO.

MULTICULTURAL MATHS

by Marjorie Gill and David Gill

Although a fact not clearly or readily recognised by everybody, England has become within the last generation a multiracial and multicultural society. In many of our urban schools we have sizable minorities of Indian, Pakistani, Caribbean, African and Greek Cypriot children, and in some schools the ethnic mix may include children from all these groups as well as many others. The schools have adjusted themselves variously to the new circumstances, making special provision, in some cases, for ESL teaching for non-English-speaking pupils; but only a few schools have appreciated the need for a thorough-going overhaul of the traditional curriculum to fashion it into a process of education for life in a multiracial society. Without wishing to embark on a lengthy justification for such curriculum reform, we confine ourselves to saying that we are entitled to expect some modest results in terms of attitude-formation as well as results in the acquisition of knowledge and skills from whatever curriculum we make up and confidently follow. If we wish to educate children for a multiracial and multicultural society, then we must ask ourselves how relevant our existing curricula are to this aim. Of equal, or possibly greater importance, we ought, as teachers, to examine our attitudes towards the ethnic minorities in this country and the assumptions we make about their cultures. It must be generally known (vide *Organisation in the Multiracial School* by H. E. R. Townsend and E. M. Brittan) that children from our ethnic minorities are, for a host of reasons, underachieving at school. Underlying some of the failure is the conviction held by many coloured children that they are despised by society at large, and sometimes by their teachers in particular. Little attention is paid, as a rule, to their own cultures. If Indian history is to be studied, then it is likely to focus on such glorious landmarks as Robert Clive or the Black Hole of Calcutta. West Indian children will find a world in the traditional school history book where white people do all the exploring, inventing, empire-building, whereas black people earn a chapter on slavery (through little fault of their own) but otherwise remain rather anonymous colonial problems until the struggle for independence in Africa throws up individual names. Black children will learn about David Livingstone's explorations of Southern Africa but not about Equiano's[1] travels in Europe.

If children find their own cultures derided or, more commonly, overlooked then they are going to feel themselves devalued as individuals. This does not mean that primary school children, or for that matter teenagers, have a precise notion of what their culture is, but they may come to wonder why all the heroes in history and fiction are white, why all the stories they read in the international language of English are by

Vishnu — a four-armed bronze figure from South India. By courtesy of the Victoria and Albert Museum.

white authors, and feel that school studies have not been designed with them in mind. It may well be that an ethnocentric curriculum reinforces that sense of inferiority or poor self-image which the majority white society has fostered in so many young coloured people.

If one grants the need for multicultural education, the question arises: what contribution could be made by Mathematics, that most abstract and universal of subjects? That school heads of department experience difficulties in the face of such a question is only too clear from the Schools Council Working Paper 50 *(Multiracial Education: need and innovation).* Heads of department were asked whether preparation for a multiracial society should be one of the aims of their syllabuses. Whereas four out of five English, Humanities, Religious Education and Home Economics departmental heads believed that such an end was desirable, only 45% of the heads of Science thought that their subject should contribute to such an aim. Furthermore, only one out of five Science/Maths departmental heads followed a syllabus which actually included items designed to prepare pupils for the multiracial aspect of society. Such observations as "science is universal" or "science does not discriminate" were fairly common responses, and there was little specific mention at all of Mathematics.[2]

The main object of this article is to suggest that despite what appears to be a highly tentative reaction on the part of teachers so far, Mathematics does have a part to play in multicultural education. Of course, in one sense, like the findings of science, the perceptions of mathematics are universal: in Karachi, Nicosia, Bradford and Hong Kong the angles of a triangle add up to 180 degrees. This should be reassuring for everybody. But that the square on the hypotenuse of a right-angled triangle is equal to the sum of the squares on the other two sides should be especially gratifying to Greek Cypriot children if they are reminded that this was the insight of a great Greek mathematician. Indeed, once we begin to look into the origins of our school mathematics we find important contributions from several cultures other than that of the Greeks which is, to some extent, acknowledged. Surely it is not beyond our wit in our teaching of decimals to a multiracial class to explain how Hindu mathematicians invented the system in the first place, how Islamic mathematicians introduced it into Europe where it was taken up by Christian scholars. Behind the decimal system universally used today lie the intellectual efforts of three different cultures. We propose, as a modest means to help create mutual respect between ethnic minorities and the white majority, and between the ethnic minorities themselves, that mathematics teachers should pay particular attention to the work of Hindu and Islamic mathematicians, as well as to the Greeks, and find opportunities to stress their importance. The following seven items merely hint at the sort of facts and problems one might include in secondary school mathematics lessons.

1 Hindu and Muslim numerals

Indian children would be particularly interested to learn that our modern numerals 1 to 9 go back to a set of numerals found inscribed on cave walls at Nasik near Bombay. They are almost 2,000 years old and look like this:

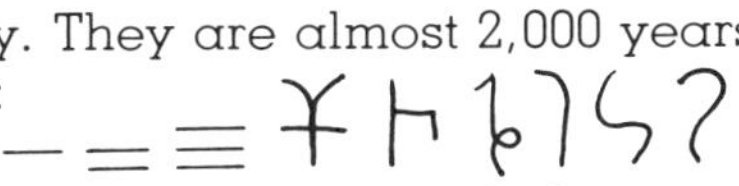

Pakistani children would no doubt be intrigued to see how the Hindu numerals developed in Eastern Islam and Western Islam, i.e. the Arabic world. Clearly it is the Hindu-Arabic numerals which lead to the European numbers with which we are familiar:

Eastern Islamic: 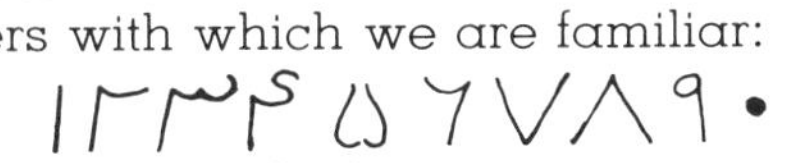

Western Islamic:
(West Arabic or Gobar) 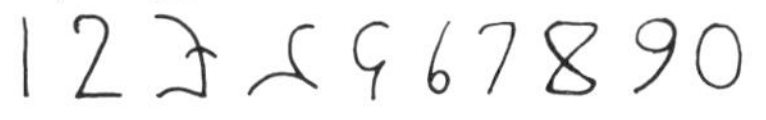

European: 1234567890

2 Much ado about Nothing

Hindu mathematicians invented the symbol for nothing, perhaps one of the greatest moves ever made in science. It enabled them to liberate themselves from the columns of the abacus and work on methods of written calculation. Originally the sign was a dot and was called "sunyabindu" (the dot marking a blank), which became in Arabic "as-sifr" or "sifr", and later in Latin "zephirum" and Italian "zero". By introducing place value and by naming powers of ten the Hindus invented the decimal system.

3 Mohammed Ben Musa Al-Khowarizmi was one of the greatest scientists assembled by the Caliph Al-Mamun to take part in the work of his academy or House of Wisdom at Baghdad in the early 9th century. As far as arithmetic is concerned, he stands as the bridge between East and West. He taught the Hindu methods of computation to the Arabs. His Arabic works were subsequently translated into Latin and the so-called Gobar numerals (Hindu-Arabic forms) became, after some initial resistance, popular in Europe. It is interesting to observe that Florentine bankers were forbidden in 1259 to use the "infidel symbols". Such religious prejudice against Islamic-cum-Hindu numerals was by no means universal in the Church. Many monks welcomed the "infidel symbols" to help them devise their calendars, and one, a certain Adelard of Bath, was so interested in multicultural education that he went to the length of disguising himself as a Muslim and in 1120 studied at the Islamic university of Cordova where he translated Arab numerical tables.

4 A Hindu rule for dividing fractions

The Hindu mathematician Mahavira (AD 850) gave the rule for dividing one fraction by another, which we still use today: Make the denominator the numerator and then multiply, e.g.:

$$\frac{3}{7}\div\frac{4}{9}=\frac{3}{7}\times\frac{9}{4}=\frac{27}{28}$$

5 Algebra

The Hindus also practised Algebra, but the name is Arabic and derives from a great work by Al-Khowarizmi (see 3), his *Al-jabr w'al-mugabala* which means "transposing negative quantities in an equation" (al-jabr) and "collecting like terms" (al-mugabala). The influence of this work on European mathematics was incalculable!

6 An Indian mathematician and some combinations and permutations

Bhascara Acharya, or Bhascara the Learned, was a famous Indian astronomer of the twelfth century. His interest in mathematics was no less acute than his interest in the stars. Here are three of his problems which could be introduced when permutations and combinations are being taught:

1. Example: In a pleasant, spacious and elegant edifice, with eight doors, constructed by a skilful architect, as a palace for the lord of the land, tell me the permutations of apertures taken one, two, three, etc. In other words, what variations can you have with one door open and the rest closed, two open and the rest shut, and so on?[3]

Statement:

8 7 6 5 4 3 2 1
1 2 3 4 5 6 7 8

Answer: The number of ways in which the doors may be opened by ones, twos, or threes, etc. is

	8	28	56	70	56	28	8	1
Doors	(1)	(2)	(3)	(4)	(5)	(6)	(7)	(8)

i.e. the combinations are

${}_8C_1,\ {}_8C_2 \ldots {}_8C_8$

e.g. ${}_8C_3=\dfrac{8\times7\times6}{3\times2}=56$

2. Example: How many are the variations of form of the god Siva by the exchange of his ten attributes held reciprocally in his several hands: namely

the rope	the trident
the elephant's hook	the bedstead
the serpent	the dagger
the tabor	the arrow
the skull	the bow?

Paramesvara, an aspect of Siva, ten-armed and mounted on a bull. By courtesy of the Victoria and Albert Museum.

Statement: Number of places — 10.
Answer: The variations of form of the god are found to be 3628800. In this example, variations means permutations in the modern sense.

$_{10}P_{10} = 10 \times 9 \times 8 \times 7 \times 6 \times 5 \times 4 \times 3 \times 2 = 3628800$

3. A simpler form of the problem concerns the Hindu god Vishnu who is traditionally depicted with four hands bearing

a mace   a lotus
a discus   and a conch

How many permutations of these symbols would be possible?
Statement: Number of places: 4.
Answer: The number of permutations would be 24.

7 A Persian lunar volvelle and almanac disc

Lancelot Hogben describes the principles on which Muslim sun-dials were constructed, and suggests how, with the aid of trigonometry you might make your own (pp. 329-332, *Mathematics for the Million*). Muslim mathematicians also designed instruments for predicting the phases of the moon. The following model, which should give excellent practice in the use of protractor and compass, is a simplified version of a Persian lunar volvelle and almanac disc which was on display at the 1976 Exhibition of Islamic Science and Technology.
The model consists of three concentric discs cut out of white card and held together by a paper stud.
The largest disc, forming the base of the instrument, is divided into 30 segments and callibrated from 1 to 30/31, as shown in the illustration.
The middle disc is likewise divided into 30 segments, and circles, radius the width of the segment, should be drawn in all 30 segments at a uniform distance (e.g. 1 cm) from the circumference. These should be marked with the phases of the moon as shown.
The top disc, a little smaller than the middle one, may be cut to include a pointer, behind which a circle should be cut out to reveal the moon-circles on the middle disc.

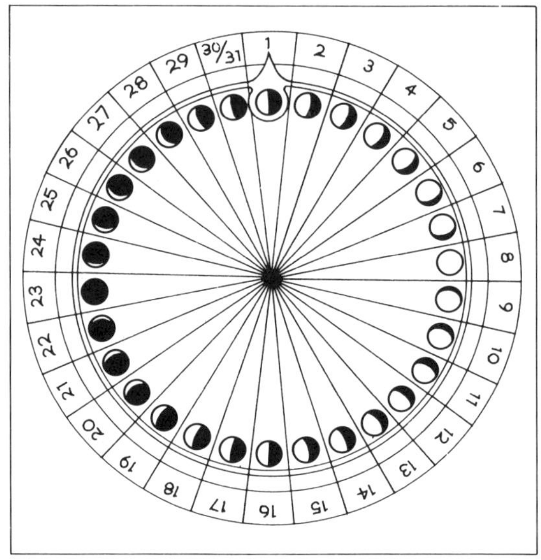

Operating Instructions
1. The top disc should be moved each day in a clockwise direction to reveal the next phase of the moon.
2. The middle disc should remain fixed, i.e. held at rest when the other discs are moved.
3. The bottom disc with the dates will normally be at rest, but will need to be moved for certain adjustments:
(a) The pointer should be set at the appropriate date initially. This may be full moon (by observation) or new moon, first quarter, etc. (taken from a diary).
(b) In the case of months with 31 days, the bottom disc will need to be moved one day in a clockwise direction at the end of the month.
(c) At the end of February, the bottom disc must be moved two days in an anticlockwise direction.
(d) The bottom disc should be moved one day in a clockwise direction approximately every two months, as the synodic month is 29.53059 days.

Notes
1. *Equiano's Travels:* his autobiography — The Interesting Narrative of the Life of Olaudah Equiano or Gustavus Vassa the African (Heinemann Educational Books Ltd.). A fascinating memoir of an 18th century Ibo abducted as a child and transported as a slave to the West Indies. He travelled widely in the Caribbean and in Europe before and after buying himself free. He later played an important part in Wilberforce's campaign to abolish slavery.
2. p.34 Schools Council Working Paper 50, *Multiracial Education: need and innovation*, Evans/Methuen Educational, 1973.
3. pp. 128, 129 *The Treasury of Mathematics.*

Bibliography
EL-SAID, I. and PARMAN, A.
Geometric Concepts in Islamic Art, World of Islam Festival Publishing Co., 1976.
HOGBEN, L.
Mathematics for the Million, Allen and Unwin, 1969.
MIDONICK, H.
The Treasury of Mathematics, Peter Owen, 1965 (also in Penguin).
MORGAN, B.
Men and Discoveries in Mathematics, John Murray, 1972.
SMELTZER, D.
Man and Number, Black, 1970.

THE STORY OF TANYA GRAM

by Audrey Bomford, Student, Didsbury School of Education, Manchester Polytechnic

Once, there was a little girl called Tanya Gram. She lived in a bungalow like this. The trouble was, she always wanted to go upstairs to bed.

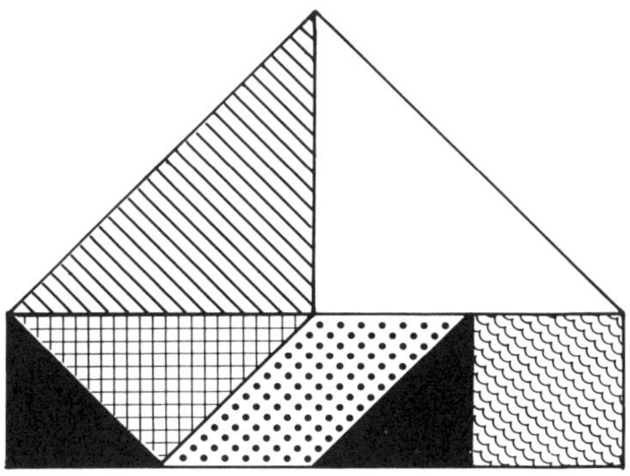

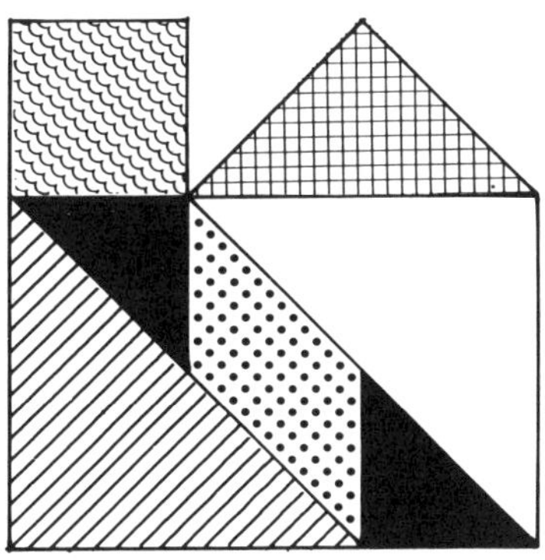

So one day she decided to pull the bungalow to pieces and make a house instead. At last there were real stairs.

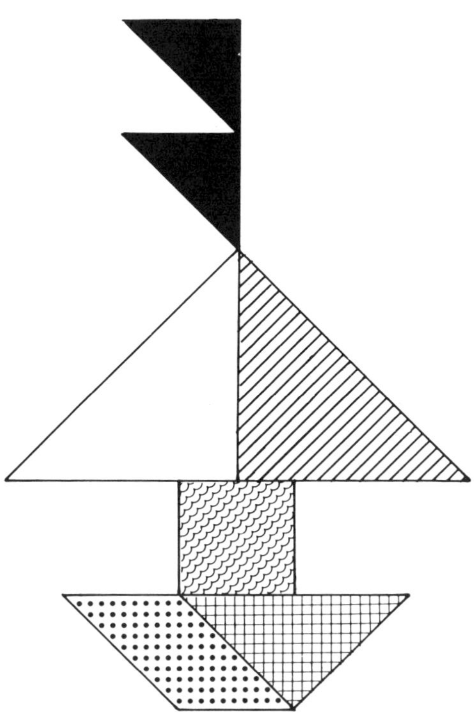

But Tanya soon became tired of living in a house, so she pulled it to pieces, built a boat with it and sailed to America.

In America she pulled the boat apart again and made a lunar module. Then she climbed in and went up on a rocket to Mars!

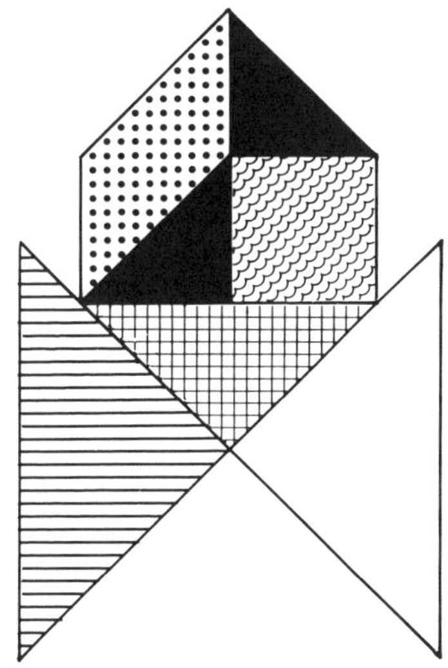

When she landed in Mars a strange man came to meet her.

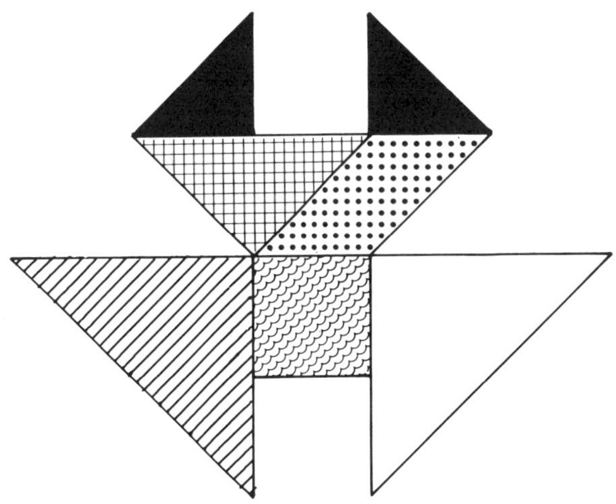

''I'm the Iggly Poggly man'', he said. ''Who are you?'' She told him who she was and that she was looking for a new home.
''I've made one out of a square'', he said. ''Come and see.''

They flew on the back of an Iggly Poggly bird to a canal and . . .

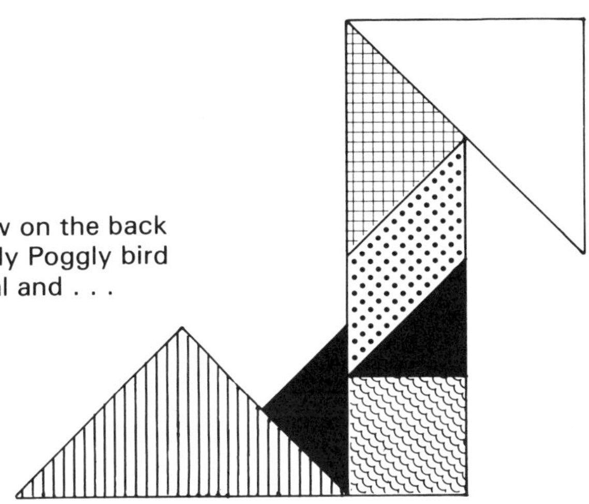

there was a lovely . . .

Niggly Poggly ark – which could go both ways at once!

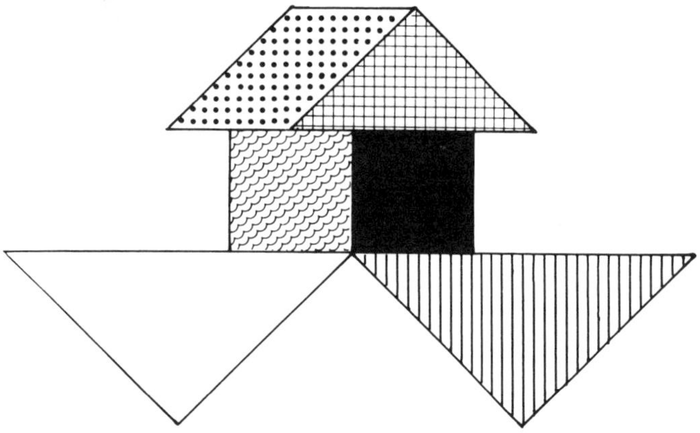

"How did you make such an exciting home?", asked Tanya.

"Well, I cut down some square Mars trees with a Mars axe, and then I fixed them together with Mars Bars!"

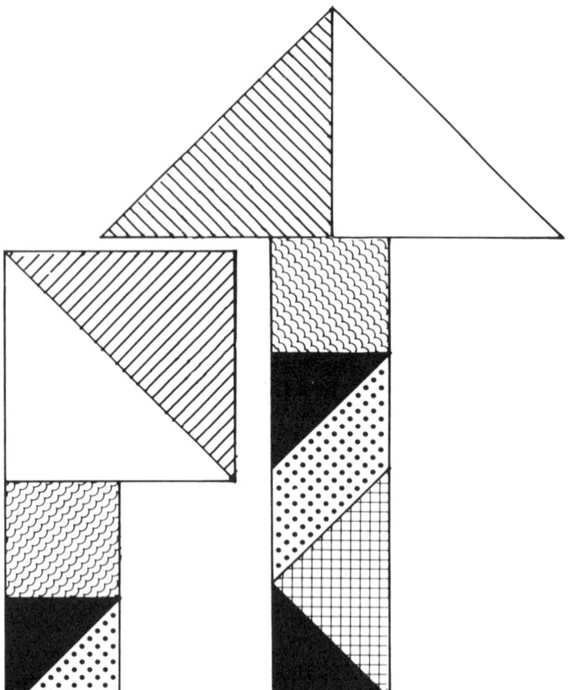

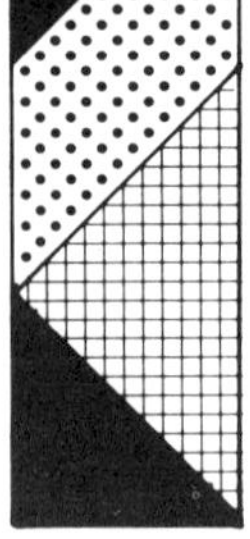

"But those aren't square trees!", cried Tanya.
"Oh, but they were before they grew into that shape", replied the Iggly Poggly man.

Just then they heard a noise. Now at the mention of Mars Bars the dreaded Riggly Poggly rhinoceros awoke. Everyone knows how dangerous he is when he is hungry so Tanya ran as fast as she could all the way back to the lunar module.

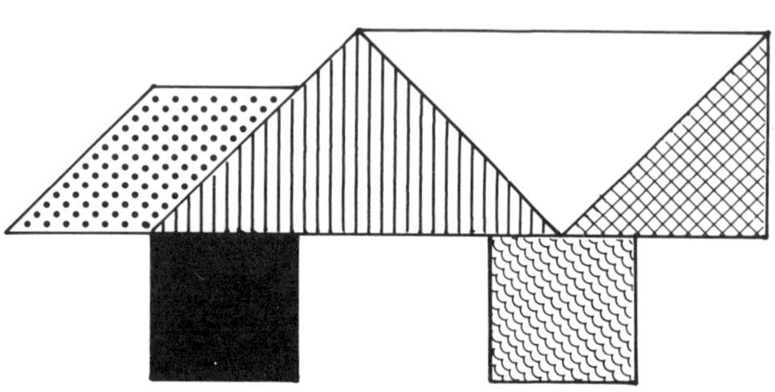

But when she got there it had turned into a . . .

rocket!

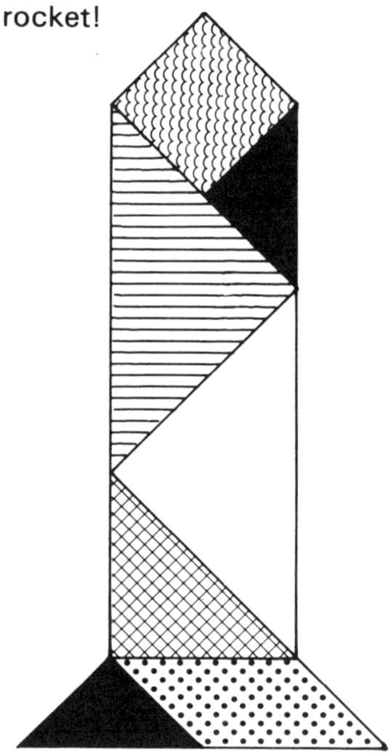

"But I can't go back home in *that!*", cried Tanya, and she pulled it to pieces and tried to make it into a module again, or a boat, or a car, or a bird, or anything that would take her home.

It simply

refused to be made into anything else!

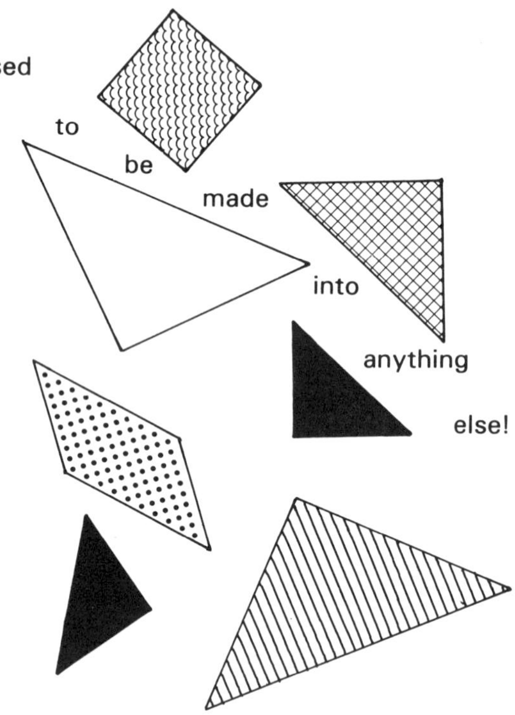

In floods of tears, Tanya threw herself down on the ground, and found she was on her own bed! She was at home, in her own bungalow, and she had been dreaming!

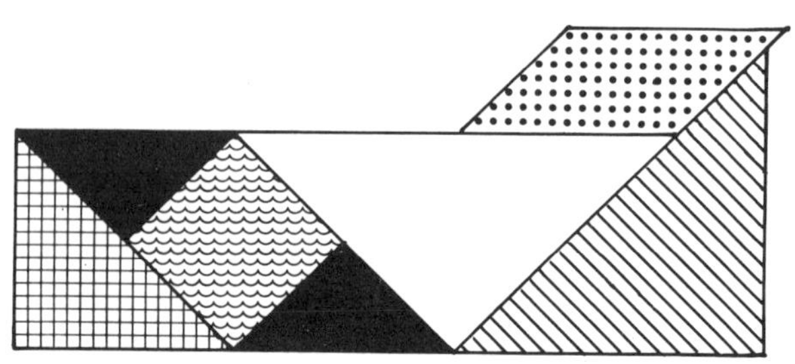

. . . and she decided she never wanted to live anywhere else ever again.

GETTING THE BOREDOM OUT OF PRACTICE

by Anita Straker, Mathematics Adviser, Wiltshire

The Great Debate has highlighted a facet of learning mathematics that experienced teachers have appreciated all along: as part of most mathematical work there will need to be both oral and written practice of those numerical skills the children have acquired. In their document (1977) *The Teaching of Mathematics in Maintained Schools in England*, HM Inspectors made clear distinctions between two kinds of numeracy:

"By algorithmic skill we mean the ability to do sums; by informed numeracy we mean the ability to know what sums to do, and when to do them, and how to use the answer. This includes the ability to avoid calculation when it is not needed, and to make short cut approximations when they suffice."

Ability in the first of these skills may well assist in obtaining good examination grades, but in the real world is of little use without the second. However, perhaps there should be emphasised a third type of numeracy:

"Confidence and alacrity in handling numbers."

This article concentrates on ways of inculcating this last skill.

Practice presented as never-ending pages of repetitive sums, or perpetual rote chanting of tables, is self-defeating. Real life is not a set of similar examples, and sums are not a model of life. For many pupils it is vital that the learning they achieve is seen by them to be relevant. Many teachers will say: "But they like doing sums." This, perhaps, is true. Sums in traditional form may be liked because children feel secure when faced with a card or page of calculations all of the same type. They know what is expected of them, and what the teacher requires. They may have acquired a few rules to apply, to enable them to get their answers right, with consequent praise from teachers, parents and others. Reading requirements are small and the thinking required is equally minimal.

Providing that the mathematical ideas are important, and few enough in number, it should be possible to present them in many different situations. This will ensure that pupils will meet these ideas frequently, and practise acquired skills regularly and often. Particularly with less-able pupils, both the concentration span and memory retention tend to be short, and their interest in whatever is presented to them will often dwindle more rapidly than with pupils of average ability.

However, for all pupils from the confident or the gifted, to the insecure or the less able, there are some generalisations to make. Any practice needs to be firstly short, so that pupils do not have time to get bored, secondly varied, so that they learn to use the same skill in as many different ways as possible, and thirdly very frequent, so that skills are not forgotten.

What to practise is not usually a problem for the teacher. It is the question of how the pupils should practise that creates difficulty. To provide for all pupils both appropriate motivation and maintenance of security in a variety of circumstances is the challenge.

One way is to try and avoid presenting them with situations where the teacher knows, and the pupils struggle to give right answers. Watching groups of children in mathematics lessons makes obvious the fact that they try to read the facial clues of the teacher, they attempt to follow the rules given to achieve what is required, in order to do things "the teacher's way". Why not use the skills they have to develop their own methods? After all, any method which works consistently can be a satisfactory one, and later, refinements can be made to improve on the time necessary to produce the same results.

Another way is to use the mundane arithmetic involved in the day-to-day chores of the school office and which might possibly be directed to some pupils. What this work may lack in interest, will almost certainly be compensated by the feelings of responsibility for something directly relevant to the life of the community.

When starting an actual lesson, mental arithmetic could well emerge within a discussion starting around almost anything, and if the pupils have suggested the topic, so much the better. What about pop record placings, football results, weather details, shopping prices and discounts, costings of student holidays, betting odds, actual newspaper headlines? (One of my favourites reads "Are your ears worth £850?", which leads to all sorts of interesting questions about composite values of parts of the body, and comparisons with compensations for injuries received for total paralysis!)

Pupils who are more motivated towards mathematics in its own right may become involved in discussing a mathematical conjecture, some statistical data, a brief piece of the history of measurement, a number puzzle or pattern, methods of Vedic Arithmetic, and so on. By the time the able pupils have tried to produce a counter example to Goldbach's unproven conjecture that every even number greater than two can be expressed as the sum of two primes, a lot of arithmetic has been done. Bowman's *Romance in Arithmetic*, or Keith Ellis's *Number Power*, contain good ideas of this type.

The Guinness Book of Records is another good source for a pupil (or a teacher) to find some numerical data about which the other pupils can pose questions. Did you know that the greatest slimming feat on record for a woman is a loss of 182 kg in 14 months, with vital statistics changed from 200-213-213 cm to 86-71-91 cm? Were you aware that the flea, which is about 2 mm long, can jump a distance of 150 times its own length, and about 130 times its own height?

Oral practice is probably more important than written practice. Real life in the adult world is more about thinking in one's head rather than on paper. In the past some ways of attempting mental arithmetic have traditionally included such things as "table tests" or "20 quick questions" as an aid to producing automatic responses. Nevertheless, varying the wording or language used as much as possible, is more likely to produce in the end pupils who are quick and accurate in the variety of situations they will meet in their everyday lives. For example, the oral wording of the question:

$$7 \times 3 \text{ or } 3 \times 7$$

could include:

What is seven times three?
What are seven threes (or three sevens)?
Find the product of seven and three.
What is seven multiplied by three?
If three children each have seven sweets, how many sweets altogether?
If I eat three meals a day for a week, how many meals have I had?
A shop makes pies in three sizes and with seven sorts of filling. How many different kinds of pies do they make?

Oral practice with certain children will almost certainly need some kind of visual reinforcement, otherwise somebody is sure to say "What was that?" or "Can you say it again?" What is important is that the calculations are attempted mentally. Answers can, of course, be written down since if group work is being done all the responses may come from one child. Another possibility is to provide each pupil with sets of cards, double width for numerals 10-90 and single width for numerals 1-9. To provide an answer the child places the appropriate units card on top of the 10s card, and holds it still on the tabletop while the teacher checks.

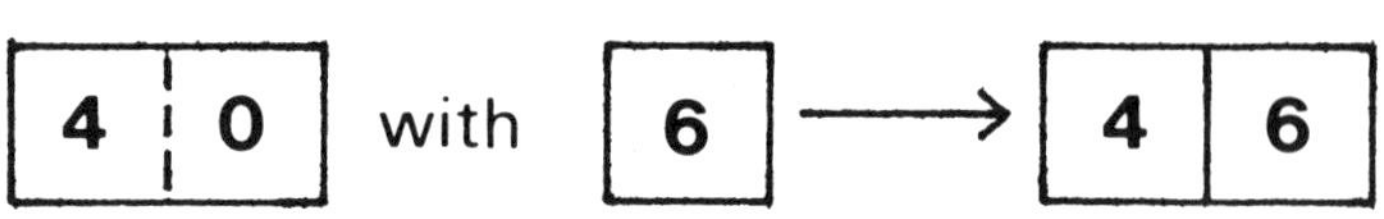

Asking a pupil who responds quickly "How did you work it out?" may help to highlight short-cut methods (including possibly some unusual ones!)

More varied oral practice can be in the form of a puzzle, eg "I'm thinking of a number. It's two threes, it's five plus one . . ." If this type of question is made very easy at first, the pupils should steadily progress to more difficult examples.

Think of a number, and let the children ask questions about it, to which only the answers "Yes" or "No" will be given, eg "Is it odd? Does it end in 3? Is it bigger than 50? . . ."

Write on the board some triples of numbers which are related by some rule, or belong to the same class, eg

$$(1, 2, 6) \qquad (2, 3, 4)$$

"What is the rule in mind?" Add a further example (0, 0, 9), and a fourth, if necessary, to encourage responses.

Let a pupil select a card from a pack of playing cards, and get everyone to write down the number with which it would total 10, or the result when multiplied by 8. (One pupil could be used as the recorder of the questions, to which answers are given at the end.)

Try a game of Fizz and Buzz: successive counting round the group with multiples of a selected number described as Fizz and numbers having the chosen figure as one of its digits as Buzz.

Allow a pupil to select a number. If it is even, divide by 2. If it is odd, multiply by 3 and add 1. Continue doing this (and in consequence a lot of simple arithmetic) to investigate what happens.

Use the numbers 1, 2, 3 and 4, and the operations $+$, $-$, $\times$ and $\div$ to form equations which total as many different numbers as possible. What about using four 2s and the four operations? What extensions are possible if, say, squared numbers are allowed? How far can successive integers using four 4s and conventional operational signs and symbols, be constructed? How many ways of forming 100, using just four identical digits, can be found?

The alternative to oral practice is written work. Here it is important to remember that if the objective of such practice is to test mastery of skills, to improve speed, or to help commit factual information from temporary to permanent memory store, then each pupil should have a high rate of success. If not, it is inappropriate for the purpose. Practice is not a technique to be used for the introduction of a topic, but should take place after an idea has been grasped.

To assist with accuracy any worksheets or workcards should look attractive. When making or designing these, it might be possible to co-opt the help of a willing colleague with an artistic flair. Key statements need to be underlined or presented in larger print. Sometimes it is possible to make the print heavier or to use a different colour. Another technique is to surround the key statements with a frame in order to bring them to a pupil's notice. This may also mean that children who have good visual memories may be helped in retaining any conclusions reached.

Similarly, the pupils themselves should be encouraged to present their work so that it looks good. Apart from anything else, setting out work well may help to avoid mistakes being made. It is possible to provide children with numbers to shuffle around on a tabletop as a preliminary to copying into books. Commercially made plastic tablets displaying numerals are available but numerals drawn on squares or circles of card are also adequate particularly for practice of the following type:

0	4	5	3
2	9	17	1
6	8	10	7

Giving them some cards and an accompanying blank sheet for answers and letting them select appropriate activities, eg multiplying each number by 7, adding 10 (or 9, or . . .), halving, totalling to 10 (or 11, or . . .). The answer sheets can be given to another pupil to mark. Similar cards can, of course, be used for class games of Bingo: "9 + 8, 18 − 10 and so on".

Using cards or tablets numbered from 1-15, complete as many equations as possible. Each number may only be used once. (There are obviously many different variations and extensions of this type.)

Making dartboards, with each of five players allowed up to six throws, and finding how each player scores exactly 100.

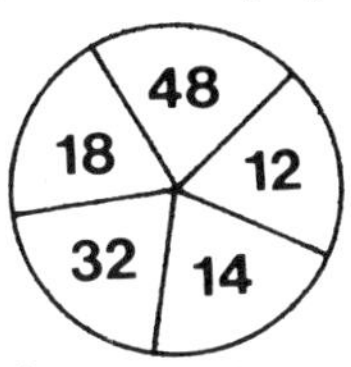

Using any three of these cards complete the following sentences:

(6) (4) (3) (2)

○ − ○ − ○ = 0	○ − ○ + ○ = 5
○ + ○ − ○ = 1	○ + ○ − ○ = 7
○ + ○ − ○ = 3	○ + ○ − ○ = 8
○ + ○ − ○ = 4	○ + ○ + ○ = 9

How many different answers could be obtained if all four of the cards above were used? Can 2 be found in this way, or 10?

Placing the first seven counting numbers so that each line has the same total:

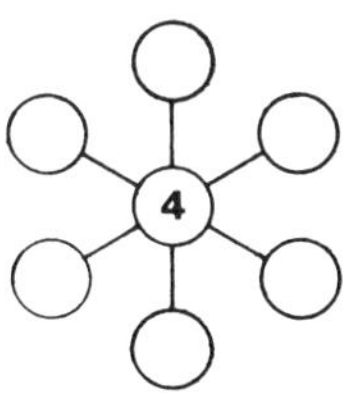

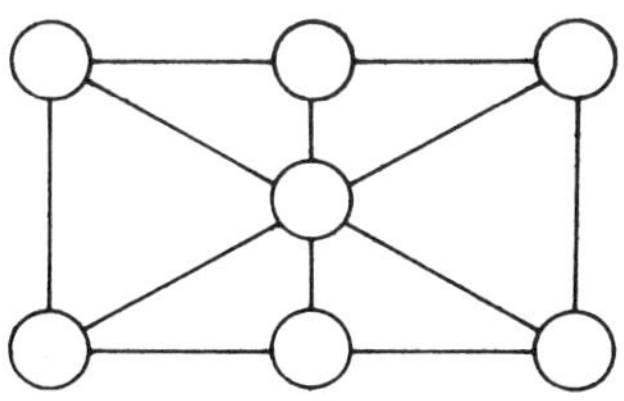

Placing 1, 2, 3, 4, 5 so that each line totals 24:

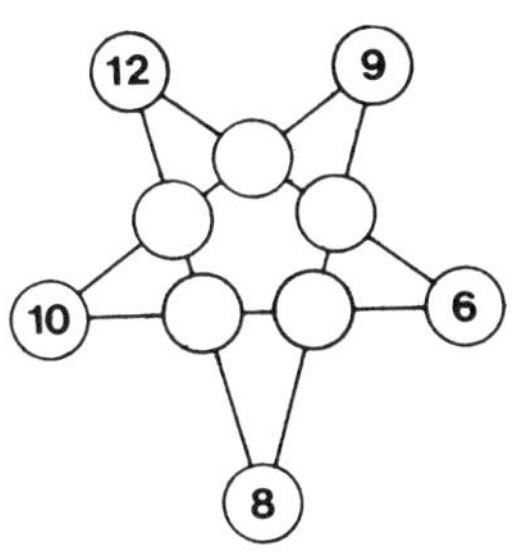

Placing the first six (or in (b), nine) counting numbers so that the total of numbers along each side of the triangle is the same:

(a)

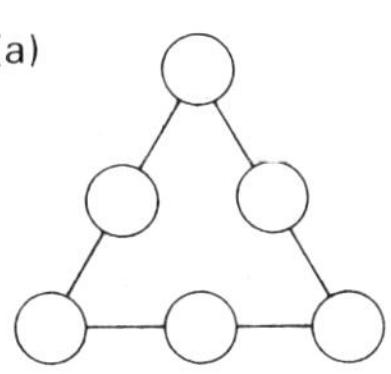

(b)

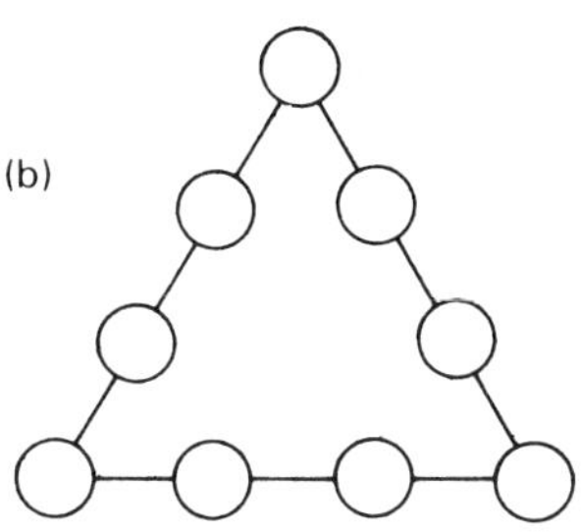

Dot-to-dot puzzles, with the points to be joined presented as simple calculations, rather than as straightforward numbers, may prove a more interesting way to practise for some pupils.

With older pupils, trying a worksheet which uses very large numbers can be a challenge, eg How would you use £250 000

if you won it on the pools?
if you wanted to buy a football team (and what percentage cut would the players get?)
to award the prizes to the last 16 players in the women's singles at Wimbledon?
to improve the wages of the number of people you employ (and would the increase be within the pay code)?

Cross-numbers, where clues are given as calculations, are easy to invent and have an in-built cross-checking to assist with accuracy. Pupils can be asked to invent their own cross-number and then provide the clues and a blank outline for a partner to complete. Magic squares also serve the same self-checking purpose.

For those able pupils (who also need to practice arithmetic), asterithms and other number puzzles or problems can be appropriate:

Find the largest power of 45 which divides 51!

In this long multiplication, find the missing digits:

```
  ***
   *9
-----
***40
 4*33
-----
*****
-----
```

In these calculations, find which digits the letters stand for:

```
  TEN          ALE
  TEN          RUM
FORTY       ------
------        WINE
SIXTY        WUWL
------       EWWE
            ------
            ERMPNE
```

Are all numbers of the form abcabc divisible by 7, 11 and 13? If so, why?

Much material similar to the above can be found in two invaluable and inexpensive paperbacks: Michael Holt's *Fun with Numbers* and *Number Games and Puzzles* by Gyles Brandreth.

But to return to the question of practice: How can we increase motivation, and thereby success, by devising circumstances which apparently put the pupils in control of their own learning? Many games allow the pupils to feel in command of all the facts, in a perfectly real and concrete situation. Sometimes this freedom from pressure and uncertainty will produce superior responses from pupils who would not otherwise have appeared capable of them.

Traditional games like Dominoes, Snakes and Ladders, Monopoly, Roulette, and card games such as Snap, Rummy, Pontoon, Cribbage, Pelmanism involve much numerical activity. In all competitive sports and games, either played actively, read about or observed, are elements of comparison, counting and measurement of length, height, distance, time and speed, necessitating manipulation of numbers.

Homemade games can be made with arrays of numbers, with pre-selected starting and stopping points, but freedom of choice as to the path taken, and used by players in turn to get nearest to a given pre-determined total: eg

(a)

−2	+9	+4	−7
+11	−6	−3	+10
−20	+16	+12	−1
+8	−9	−7	+5

(b)

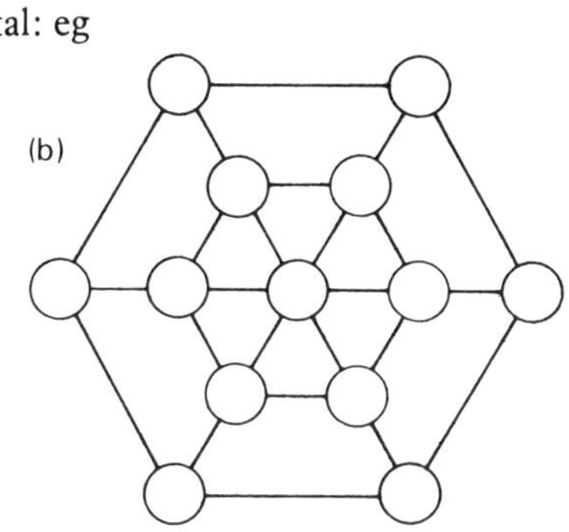

Commercially produced games abound. Catena from Arnold for practice in multiplication tables, Quick Tens (or Teens) from Galts for addition and subtraction facts, Mathemax from Hope and Equable from Invicta for progressive number situations, are just some of the many good ones.

Another useful piece of equipment is a pre-programmed calculator from Texas Intruments called "The Little Professor". The pupil can select which of the four operations he wishes to practise, and one of four levels of difficulty. On pressing the GO button, 10 questions are displayed in turn for the pupil to type in an answer. (If it is incorrect, the calculator repeats the question.) At the end. the pupil's score out of 10 flashes on the display panel. A more sophisticated version, Dataman, has in addition box arithmetic, a timing device, and a choice of several games.

Apart from the games described here, many of the activities described in this article can be adapted so that the pupils are apparently in control of their practice, rather than the teacher. There are many ways of doing this, for example:

Getting pupils, providing they can do the calculations, to set some for others. (This can, in fact, be quite challenging. It is usually more difficult to invent the questions than actually to do them, as any teacher knows!)

Giving out completed, teacher-made worksheets for pupils to mark and correct.

Making up a number puzzle and then giving it to a partner to solve.

Letting them mark their own work by using an electronic calculator.

Writing a number (chosen by the pupils) on the blackboard, and having 2 min, say, to write down as many sums as possible in that time having an answer equalling the number on the board.

Encouraging them to be involved by making them think they are choosing numbers (perhaps by rolling one or more dice or selecting them from a pack of playing cards).

The ideas in this article have been gathered from many sources and are by no means exhaustive. But perhaps if some of the approaches suggested here were incorporated into part of the majority of mathematics lessons, it would help to meet the criticisms currently being levied that our pupils are no longer numerate.

And finally, as an end of term change, it would always be possible to chalk 40 sums on the blackboard!

References

1. Margaret E. Bowman (1950) *Romance in Arithmetic*, ULP.
2. Keith Ellis (1977) *Number Power*, Heinemann.
3. Norris McWhirter *Guinness Book of Records*, Guinness Superlatives Ltd.
4. Michael Holt (1976) *Fun with Numbers*, Piccolo.
5. Gyles Brandreth (1976) *Number Games and Puzzles*, Carousel.

SOCIABLE CUBES

by Donald Thornley

Donald Thornley is a student teacher on the Maths Induction Course devised by the Inner London Education Authority to enable mature entrants to the teaching profession to add their own particular strengths to the maths departments of London schools. He is one of 20 recruits from industry who joined the Authority's teaching force in January to undertake a two term course at Chelsea College Centre for Science Education. The course is serviced by Chelsea College and the ILEA inspectorate and advisory staff. It has been valuable in strengthening relationships between the maths teachers in the schools and the supporting "maths team" under the staff inspector Laurie Buxton through their joint responsibility in the training of these teachers. Course organiser is Miss R. F. Gibbons.

1. Sociable Cubes get their name because they readily join together to form colonies.

The Cubes that we have here like to form colonies of not more or less than eight cubes.

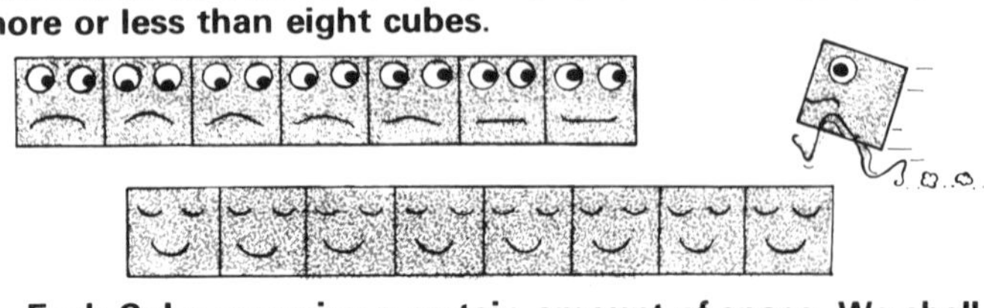

2. Each Cube occupies a certain amount of space. We shall call this amount of space "one unit volume" or one "unit".

3. The status of a colony depends on its size. What is the biggest volume of space that a colony can occupy?

4. No matter how hard they try no colony has ever managed to occupy more than eight unit volumes.

5. However, Sociable Cubes have discovered an interesting fact. In winter Sociable Cubes need to wear a coat that completely covers them.

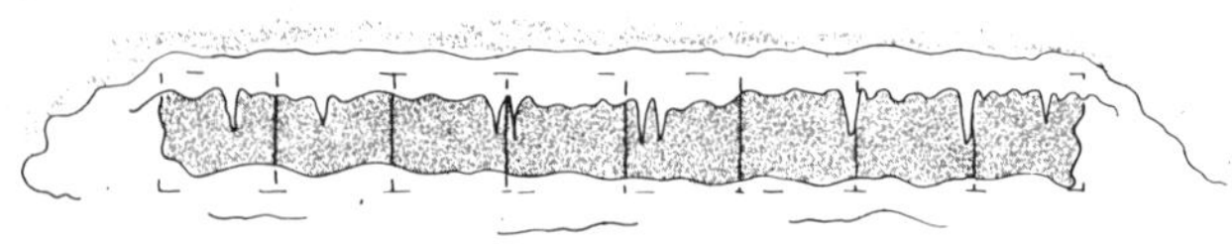

6. They have discovered from their tailor that some colonies need much more material to cover them than other colonies. The colonies needing the most material have most "face".

The area of the coat is of course the surface area of the colony.

7. What is the largest possible surface area for eight cubes?

What is the smallest possible surface area for eight cubes?

CHILDREN'S LANGUAGE AND MATHEMATICS

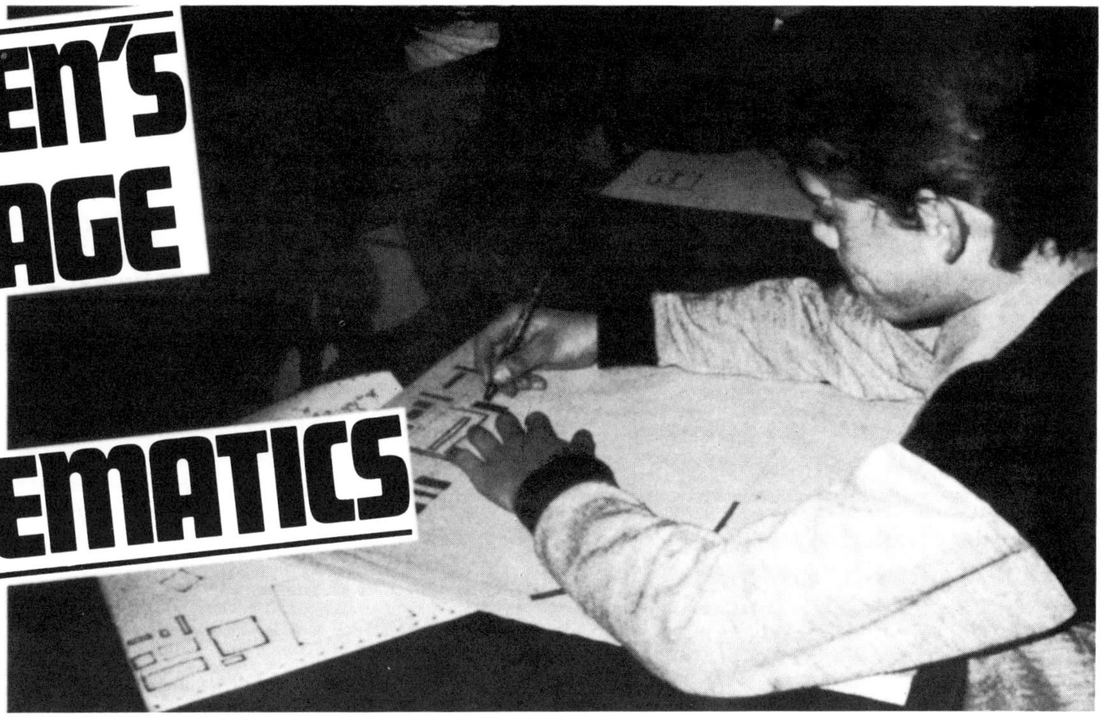

REWRITING THE LESSON SCRIPT

by Tom Brissenden,
University College of Swansea

There can surely be few teachers who by now are not familiar with the famous Cockcroft paragraph 243, which demands amongst other things "discussion between teacher and pupils and between pupils themselves" as an essential feature of mathematics lessons at both primary and secondary level. The phrase gives official sanction to something that has been increasingly recognised by teachers, educators and researchers in recent years, namely the importance of talk in the learning of the subject. Recognising and recommending are all very well, but the practical problems of developing mathematical discussion in the classroom are considerable. It is this aspect with which this series of articles is concerned.

I have been fortunate enough to be able to exchange ideas with several other people working in this field. We have got to know about each other's approaches, which seem to us to complement and extend each other. Our interests are practical, classroom ones, research-oriented only to the extent of drawing on work that seems to us important and useful. Our experiences during a teachers' course in Swansea in 1984 were the genesis of this article. We believe our suggestions are practical ones that will help teachers to develop their own discussion lessons. Each of us expects to have training materials available in 1985, so readers interested in following-up in more detail should have no difficulty.

I begin our series by describing the development of my own approach. One way of summarising it is to liken the classroom situation to a play with a familiar plot, so that the participants fall easily into their accustomed roles. The traditional "mathematical plot" seems to leave little or no scope for discussion, so that what we arc about is nothing less than "rewriting the script", for everyone concerned. In order to do this, we shall have to consider three aspects of our role, namely:

(a) **the planning and preparation of the mathematical situations we use;**
(b) **the choice of appropriate lesson patterns;**
(c) **control of the classroom interaction so as to obtain the two forms of talk mentioned by Cockcroft.**

Rewriting the Traditional Script

Imagine yourself for a moment as a pupil in a typical mathematics lesson — not a difficult task, which my student teachers can role-play "off-the-cuff" with no trouble at all. "What are seven nines?" asks your teacher. He or she wants you to put your hand up, you know, so you expect to be told off if you call out. You shoot it up along with twenty others because you think you know the answer. That idiot Jim puts his up, hoping he won't be asked, but the teacher is too smart and picks him first. Jim has to sweat out a dreadful pause, then the teacher takes time out to comment on his behaviour in unflattering terms. Quite a few hands drop down during this performance, so you wave your arm about hopefully. No luck again — it's bright girl Mary who gets the limelight. "Well done, Mary!" says the teacher, "Now what are six times eight?" You decide to sit this one out, but quite a few people call out so you join in anyway. "Forty-eight!" "Well, yes", says the teacher, "but I told you before — put your hand up if you know the answer — don't call out." You know the answer to the next question but up shoots Mary's hand so you opt out again. Even if you are picked on, it won't be so bad as with Jim — and so the lesson proceeds.

Exaggerated perhaps — but with a familiar ring? We have all spent years in the classroom, learning our pupil roles in this particular play very thoroughly. They interlock neatly with the teacher's role to produce an expected pattern, sometimes described as "question and answer". There is very little about it that has anything to do with discussion, and if the people involved are to break out of it, they have to learn very different roles.

Many people I speak with find it hard to imagine how there *can* be any discussion about mathematics. "Surely there's only one right answer", they say, "so either you know it or else you don't!" Here is our first huge obstacle in people's own minds — a complete misunderstanding of what mathematics is about. If you don't believe something is even possible, you are not likely to seek ways of achieving it! Moreover, there is some foundation for this widespread belief. If I ask a question in mathematics, I am quite likely to have a good idea of what makes a reasonable answer — what I call an "ideal response". It may be very hard for me to avoid matching pupils' replies to this ideal response and so I have a constant temptation to comment on *everything* they say. My comments will usually take the form of an evaluation which makes it clear whether I accept the response as "right" or reject it as "wrong". My pupils have learnt that this is part of my traditional role and expect me to do it, just as they expect to ask the questions, of course. So a clear pattern emerges in our traditional drama entitled "A Slice of Classroom Life". It looks like this —

TEACHER ASKS QUESTION →
PUPIL RESPONDS →
TEACHER EVALUATES STUDENT'S RESPONSE

I call this the "three-term sequence" — it recurs over and over again, so much so that I was able to construct a flow chart for it!

Discussion is not at all like this — there isn't someone whose job it is to comment on what everyone else says. If that happens, the others are liable to shut up and let the "dominant person" get on with it — the way some head teachers run their staff meetings, perhaps. Sociologists have a well-established body of findings about factors which affect discussion — group size, for instance — and about techniques for establishing and maintaining it. Do teachers know much about this? If they do, can they accept that those techniques could be employed in the teaching of mathematics?

These are the questions I found myself asking in reflecting on my own performance. I was greatly helped by observing others, including some excellent models which showed that "it can be done". I became conscious of aspects in teaching of which I had previously been completely unaware. Was I really listening to what the pupils said, for instance, or just looking for what I already had in mind? What could I learn from the mistakes my pupils made? What was really happening when I seemed to be doing a lot of the mathematics myself? It seems to me that the only real way to effect changes in one's teaching is through increasing self-awareness such as I have described. I will make some suggestions about this later, but first let me offer some examples of children talking about mathematics. In all cases the teacher is myself.

Planning Effective Mathematical Situations

Three ten-year-olds, Joanne, Kathryn and Martyn, are seated round a table with me. They are playing a "cube game" using 1 cm square lattice paper and a supply of 1 cm yellow and blue cubes. Each child in turn constructs a shape; the other two check whether it is different from all the shapes previously constructed *and have to say why*, with the builder justifying if necessary. The rules for making shapes (in this game) were: a base of five yellow cubes must be placed inside a 3×2 rectangle drawn on the lattice, then two blue cubes placed on the base, all the cubes having to be "fitted together properly". After a short practice session, Joanne made the first shape shown in the diagram below, and Kathryn completes the second. We follow the discussion through the first round.

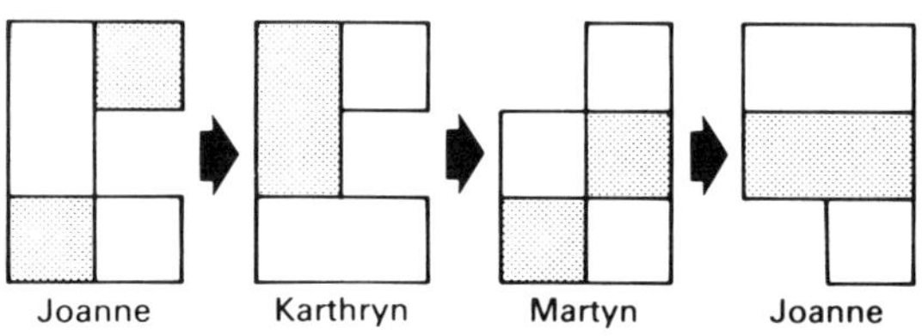

"Ah, now, Joanne, is it different from yours?" (Kathryn's shape.)
"Ye-es."
"In what way?"
"Cos it's got the same shape in yellow as me but the two blues are in a different place."
"Are you happy about that Martyn? Kathryn's is different? ... Now it's your turn, I think, isn't it?"
(Martyn builds his shape.)
"Hmm — is it different from yours, Kathryn?"
"Yes."
"How's it different?"
"There are still five yellow ones but then he's left the space in a different place and, um, my two blue ones are touching rather than leaving a space in the middle."
"Joanne — is it different from yours?"
"Yes."
"In what way?"
"Cos, erm, his yellows are — he did five as well but, erm, I've — I've left the space in a different place and the two blues, erm ... huh, erm — ..."
"There's something different about the blues, isn't there? Now Kathryn's are touching, aren't they, her two blues? ... Now yours are apart like Martyn's but they're a bit different, aren't they?"
"Yes ..."
"How are they different?" (Short silence.)
"You think yours are placed differently from Joanne's, Martyn? Would you like to say then?"
"Yes, mine are ... straight to each other."
"And what about Joanne's?"
"They're sort of more diagonal." ("*Yes*", from both the others.)
"All right ... is it Joanne's turn?" (Joanne builds the fourth shape.)
"O.K., now is that different from yours, Kathryn?"
"Yes ... Joanne's left the first square missing ... and um, the two blue ones are placed in a different way from mine."
"Now what about yours, Martyn?"
"Mine's different from hers because the shapes — the yellow shapes are the same but the other way round, the blues are different."
"We've got a little problem here — are we going to count this base — let's just look at the bases for the moment — are we going to count that as really the same as Martyn's or not?"
"I don't think it is." (Decisive agreement from Kathryn and Martyn as Joanne says this.) *"If I turned it round then the square missing would be there and not by there."*
"Ah, that's a good point, isn't it? Shall we turn it round? What do you mean — like this, Joanne, there?"

Remember that the children were seated round a table with their own shapes on a sheet of paper in front of them, so their view was *not* as in the diagram. The discussion continued in this manner through seven rounds and twenty-one shapes constructed (which the teacher praised as a *group* achievement). They developed close involvement as the activity proceeded, with great excitement at several stages.

This extract illustrates one of the most important factors in getting discussion — you must have an "effective mathematical situation" in which the children can work. In particular it must be effective in enabling the children to make their *own* judgements about how they are getting on, and minimising the amount of mathematics the teacher needs to do. Obviously the concrete form of the present situation facilitates this. The game format produces the talk, through the checking of a player's shape by the others. This idea is useful to transform many other situations into discussion ones. Janet Duffin and Nick James will be describing other examples of effective situations in their later articles.

You will notice that the teacher's presence is essential on this occasion to get the children talking. I have to "run the show", not doing much mathematics perhaps, but acting as leader and keeping things going. The children's initial one-word replies are only expanded by my prompting. Constructive discussion does not just happen — children have to learn it like anything else. In time they do, and take over many aspects like asking questions — "How do you get that?", "Show us what you mean", and so on. The children in this group had not yet developed such skills. So my "procedural role", as I call it, is quite a prominent one. There are some interesting mathematical points, which are not my main concern here. Note how Joanne steals the phrase "left the space in a different place" from the more articulate Kathryn, and how the group unites on two occasions later on, particularly with Joanne about turning her shape round. Let's just hear a later round of this game, in which I have to explain my use of the word "scope". There are thirteen shapes on the table by now, so life has got complicated for the players. Martyn makes an innovation, with the first shape shown in the next diagram.

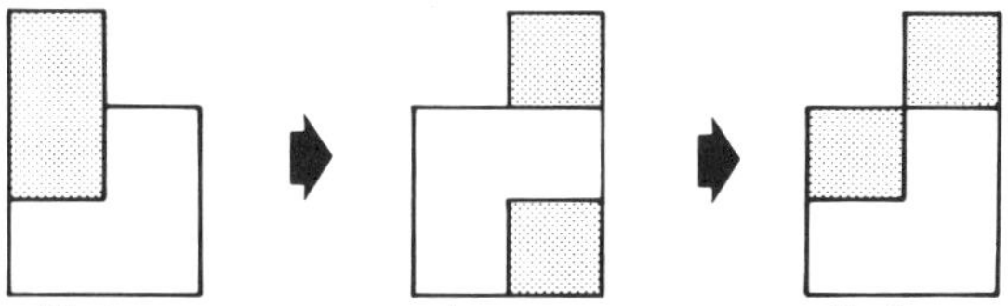

"Ah — yes! Nobody's thought of that yet, have they? There's cunning for you! You claim that's different, do you Martyn? In what way?"

"Cos my top blue is on the edge — the very edge."

"It's on the cube that's sticking out, isn't it? Yes — nobody's thought of that yet, have they? ... Joanne, come on then —"

Joanne makes the next shape, with a little prompt.

"Ah — you could try Martyn's idea, couldn't you ... how about putting a blue on the one that's sticking out and then seeing what we can do — ah, yes that's clever! Is that different from yours, Martyn?"

"Cos her base is the other way round ... and the blues are apart."

"Yes, I think we've got some scope here — right, Kathryn —"

"What's 'scope'?"

"Erm — more opportunities — more possibilities — more scope."

"Right — I didn't know what it meant." (Kathryn laughs.)

"That's what it means — Martyn's idea there has given us quite a lot more scope, I think because I think we've got a few chances to make some more haven't we, by putting our blue one on the one that's sticking out?"

Kathryn makes the next shape shown.

"Yes, I like that — Kathryn's borrowed your idea as well, Martyn. Yours is different from Martyn's first one, then, Kathryn — in what way?"

"... the yellow — one of the yellow blocks is in a different place by there and my two blues are diagonal, Martyn's aren't."

"All right — now, Joanne's got one a bit like that — is yours different from Joanne's?"

"Yes."

"Her yellow base looks the same as yours doesn't it? ... What about your blues then?"

"Joanne's blues aren't touching — mine are — on the corners."

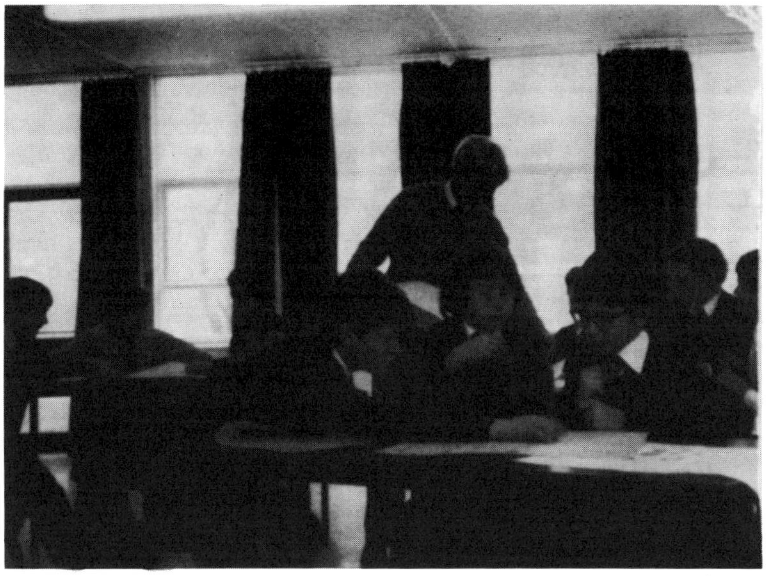

Lesson Patterns Which Generate Discussion

Getting the children together in small groups, as in the session just described, ensures participation. Making the mathematical situation into a game is a good way of getting talk going, and making sure it is about mathematics! There must be *definite tasks or goals*. There are plenty of such ideas for games about, particularly in the newer maths schemes. For the group size, I prefer three children, but there is plenty of scope for variation and experimentation in this respect. Our long-term aim is to get the children *working together* in these small groups, and *talking about mathematics* — not just sitting at the same table or indulging in social chatter. They have to learn to be independent in this respect, since the teacher obviously cannot be with all groups at once. Nick James will be describing ways of organising such group work in his later article. But the teacher often wants to work with the class as a whole. My next extracts show how this might be organised, and some techniques the teacher can use. They are from a recording made of a demonstration lesson I did in the Swansea course.

My class of eleven-year-olds was seated as in the photograph A. During the previous lesson, I found that the children used the words "square" and "rectangle" in an exclusive sense, and had met the word "quadrilateral". My aim in this lesson was to talk this through towards my own usage. I started with a warming-up session of oral arithmetic, in which I used a variant of an idea suggested by Nick James, instead of children putting their hands up.

"Now I'm going to give you a starter this time and then I want you to supply some sums to go with it. Now when the starter's twelve, then one answer will be four times three. Now when I give the starter I just want you to put your thumb up when you've thought of a sum to go with it — and I'll spot you — just like that, you see — 'I've got one'. Now we'll start with twelve, shall we? Think of a sum that goes with — Paul?"

"Three fours."

"Two sixes."

"Richard?"

"Three fours."

"Can we get away from times? ... Martyn?"

"Twelve ones."

"Jan?"

"Six ... and six."

"Natalie?"

"Eleven plus one."

"Ten plus two."

And so on, with subtraction being brought in later. The next starter is "Tell me sums like thirteen take away eight", again with a *variety* of responses which follow that pattern (23 − 8, 43 − 8, etc.). Next I use an incident from a discussion around a micro programme, and ask the groups to discuss together how they would explain "multiplying by nought" to a younger child. After a few minutes discussion, the class is brought together and spokespersons from each group report on the work that took place. During the group session the teacher takes the opportunity to circulate amongst the groups, listening and intervening. This pattern is used during the next phase of the lesson, in which the main theme is developed.

"All right, that's something we'll have to come back to — I want to take up something we were doing last time. You remember we were drawing all those shapes — we'll get them out in a minute, but before that I seem to remember that somebody started talking about quadrilaterals and I think it was you, wasn't it, Mark? ... Now, all this business about quadrilaterals — what is a quadrilateral?"
"It's a shape with four sides."
"And I think somebody said there that they had to be straight, didn't they, when you were talking about it? ... All right, now Anthony, could you give each group one of those sheets of paper please ... Got your pens? ... What I want you to do is to take those pens and your paper and talk together and draw me some quadrilaterals but — those quadrilaterals must have all their angles ninety degrees ..."

The lesson switches to group working as indicated above, while I circulate. After a time, judging that enough diagrams are drawn and using impressions gained from my conversations within the groups, I draw the class together.

"I want you to hold up your paper so that we can all have a look and see what you did ... O.K., put the papers down again. You've noticed something I expect, haven't you? Now let's go round again — who's going to talk for this group this time? I've forgotten, was it Kathryn — Anthony — yes?"
"We only found some squares and some rectangles."
"That's all you were able to find, was it?"
"Yes."
"We did small rectangles and squares," interjects Kathryn.
"Ah, you could draw them different sizes, could you?"

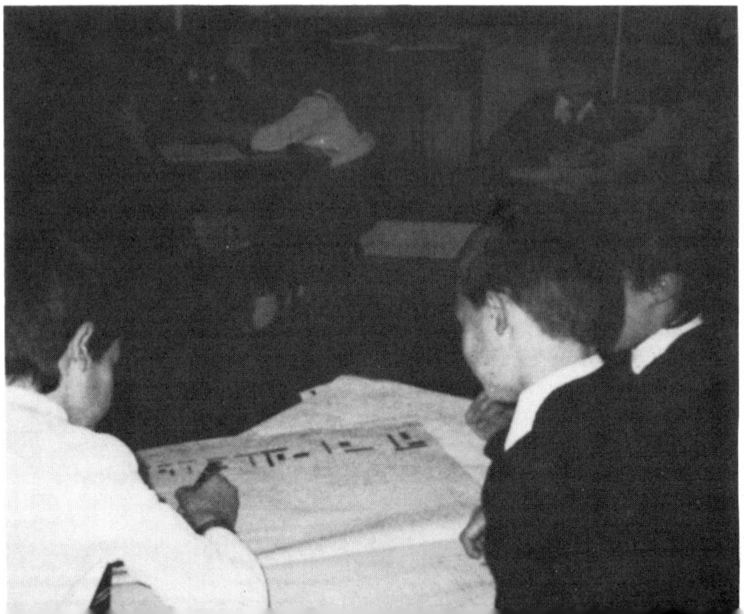

"Yes."
"Hmm, that's interesting. What about your group now? You did the talking last time, Matthew, didn't you? Mark? You started all this with this quadrilateral business, now what about your group? ... Could you talk to the class, by the way, and not to me?"
"We found the same as their group did but we found out if you put this square at a different angle it looks like a rect- a diamond."
"Can you just explain that a bit more? Do you want to hold that up and point or draw another one, Mark? ... Which one was that? Was it in your first drawings or later ones?"
"This was one square — and these four — and then we had one like this —"
"I don't think any other group did this, did they?"
"... we ... out the diamond shape was still a square."
"The diamond shape was still a square. O.K., thank you very much, Mark. What about your group? You spoke last time — O.K., Anna — what did your group find out? ... Are we all listening, please?"
Anna is a very shy pupil and "freezes".
"The same." (She whispers almost inaudibly.)
"Well, no, come on, you've got to say." (Pause.)
"Anna ..." (I move up to Anna and speak quietly and gently.) ***"What did you find out, Anna — can you point to what you found out? ... Now tell me about these shapes, Anna, please ... How many did you find?"*** (Softly to Anna only.)
"Five."
"Five, was it?" ... (I draw Anna into a short conversation with me.) ***"What kinds of shape was it?"***
"Diamonds, squares and rectangles."
"Can you all hear, by the way, what we're saying here? Can you speak a bit more loudly, Anna?"
"We found rectangles." (She speaks more clearly and audibly.)
"You found rectangles, did you?"
"Squares and diamonds."
"Squares and diamonds? It sounds a little bit like this group did. Which one was your diamond, Anna? Can you show us the one you had? ... It's that one there, is it? Can you just show it so that Rachel — sorry, Jan — O.K., erm, who spoke for your group last time? ... You going to go this time, Richard?"
"Yes. We found ten quadrilaterals and also found that it doesn't really matter how long the sides are they have all got ninety degrees."
"Hmm. What kind of shapes did you find?"
"We found roughly the same as Mark's group over there — we found the diamonds, rectangles and squares."
"Hmm, people have talked a lot about these diamonds, haven't they? But you said something interesting, didn't you Mark, about that? What did you say? Can you remind us?"
"It's a diamond shape but — it looks like a diamond shape but it's still a square."
"What do you think about that, Richard?"
"Well, I've been sticking to it about forty-five degrees and looking at it and it's just like a square."
"We'll have to decide what we're going to do about that, won't we, in a minute? Let's just find out what the other groups had to say — Jan, would you like to say?"

I have not the space to report how the lesson continued in full, but the conclusions emerging from this first sharing interlude were that my instruction had forced them to draw diamonds, squares, rectangles and oblongs, that squares were a kind of diamond and that there was another kind of diamond. Mark made a lengthy contribution which included "... If you move the square to its side it looks like a diamond but all its sides are forty-five degrees and there's another diamond turned on its side to look like forty-five degrees". I forced the issue by saying "All those shapes that you've been drawing for me *I call rectangles.* Now, you've been drawing two kinds of rectangles. Now ... two minutes — *what do I mean by that?*"

A buzz of group discussion ensued, in the same manner as before, ended when I said "Now what have we decided?" and called various (different) people to report for their groups. The crucial statement, which caused general amusement, came from Ken — "There are two kinds of rectangle a square, and er — and an oblong!" Ken's tone of reluctant apology was related to an ongoing joke between the class and myself about "oblong" being, in their phrase, "a baby word". In the final group phase the class made drawings of "an oblong rectangle" and "a square rectangle", before resuming the activity of the previous lesson.

Controlling Classroom Interaction to Generate Discussion

For my mathematical situation in that lesson I drew on experience which I knew was available to the children. The actual materials were quite simple — large sheets of computer paper for each group and coloured pencils. Having just a single large sheet to work on is another effective way of making a group of children share their ideas. The way in which the problem develops is signposted by my opening demand — "Draw me some quadrilaterals *but* — those quadrilaterals must have all their angles ninety degrees ..."

Each time I ask such a key question, you notice the class have a "thinking opportunity" — a period for individual work and discussion in groups. During these periods, I can circulate, so that when I bring the groups together I already have a good idea of the kinds of responses my demand has provoked. Moreover, the children have already been able to talk together about their ideas and difficulties. So I have gone a long way towards breaking out of the "three-term sequence" I mentioned earlier, because the children will *expect this opportunity* instead of having to compete immediately for my attention. My lesson pattern looks like this:

Whole class teacher-posed question → **Individual work and small group discussion** → **Whole class sharing of ideas** → **and so on.**

Each individual/group phase lasts several minutes and generates ideas for the next whole-class sharing phase. Here the teacher thinks of the class as a *small number of collaborating groups* instead of a large number of individuals. The seating is important — see photograph A. I use an "open horseshoes" arrangement. That is, each group of three face one another so as to interact, while the "horseshoes" so created all face some central arena for sharing, such as a blackboard, overhead projector or piece of apparatus. A blackboard is good for allowing ideas from different groups to be written up and compared simultaneously, while a projector roll saves them for viewing in succession. A projector is also effective for displaying actions in silhouette e.g. counter arrangements, rod or shape patterns or constructions with instruments.

What I have said so far should help you to rewrite your own lesson script, but the teacher needs to think carefully about his or her own role in "the play" and how to alter it. This can be more difficult — much of what I have just described relates to planning, now we have to become aware of our own actions during the lesson and try to modify them. I would sum it up by saying you should resolve never NEVER to evaluate your pupils' responses — to try to *cut out* the third term in that notorious three-term sequence. Children are not then inhibited by what they feel to be the teacher's judgement of them. The full power of the mathematical situation and the lesson pattern can then operate in an atmosphere which encourages both the tentative offering and the development to more formal discourse when this becomes important. I have tried to exemplify what I mean in the extracts quoted, with what success you can judge for yourself (I would make some important criticisms of my own performance). To develop these skills and make progress requires self-awareness and reflection on one's own lessons — a process in which tape recordings can be helpful and salutary.

To some people, at least in the context of the mathematics lesson, this suggestion of "cutting out the third term" seems shattering. Pupils also, remember, will have to adapt themselves to the new role in which they have been cast by you. ("Why doesn't he say whether I'm right or not?" may be the almost-visible reaction in certain cases.) No longer mere "spear carriers", they have leading roles in our new play, "The Problem-Solving Adventure". There are various teaching skills involved. One of the most important is listening carefully to what children actually say and mean, *not* whether their words approximate to your own "ideal response". How would *you* have reacted to some of the children's statements in my extract? Select some and answer honestly! There are various techniques — drawing out, linking, deflecting and pigeon-holing the main ones — which can be employed in handling discussion. I call this the "procedural role" of the teacher to distinguish it from his or her "mathematical role", which should consist mainly of questions.

I do not advocate "group leaders" because *all* children need to develop their powers of discussion to the greatest extent they can. So everyone, however shy or inarticulate, must take their turn at presenting ideas in public — that's why I insist with Anna, in the above extract. Would you? Other children can be supportive, *just like the teacher*, who needs to be firm about them listening to and talking to each other — you can spot the relevant moments in the extracts. I try to draw out Anna and ensure that the class is, firstly silent, and later on, that our conversation is public not private. I would quote in support of this approach something I wrote in 1980:

> "It can be argued that in some ways the classroom is like a jungle, a competitive situation in which the stronger survive at the expense of the weak, in which some actively seek and gain attention while others as actively avoid it ... Some of the failures of discussion in this situation may be because it is seen as another form of competition offering yet more opportunities to *those who already possess the skills to take advantage of it.*"

I hope I have now said enough to help you make a beginning, although of course there is an enormous amount of fascinating ideas to be considered. Discussion is not some kind of classroom anarchy, nor will it happen magically just because the teacher wants it to! Careful planning, increased self-awareness and reflection on one's own role are the essential ingredients. We shall be trying to draw the various threads together in the later article of the series, so at this stage I shall only say that I would be interested to hear from anyone trying out these or similar ideas, with examples and suggestions. My address is the Department of Education, University College of Swansea, Hendrefoilan, Swansea SA2 7NB.

3 Secondary Mathematics
by John Bradshaw

This tiny selection from the wealth of accumulated experience contained within the volumes of *Mathematics in School* reflects to some extent the interests and enthusiasms of teachers of secondary mathematics over the past seventeen years. A number of themes recur at regular intervals — a hardy annual being the *Four 4s problem* wherein it is required to generate as many numbers as possible using the digit 4 exactly four times, with whatever symbols desired. This problem was described in September 1980 and solutions for the first 50 numbers given, soon prompting another reader to send in pupils' solutions from 51 to 100. A similar article in January 1985 reminded Mervyn Hughes of a variation using only the digits 1,2,3 and 4 and he listed 50 solutions. However he also asked if there was a complete set in which the digits retained their numerical order — so far the editors have received no such solution. . .

Each January my school holds a competition in a similar vein, the aim being to obtain the numbers from 1 to 100 using the year's digits, preferably in order. Contributions come from pupils from first to sixth form but we rarely manage to complete the list — unlike Anne Williamson who provided a complete listing for 1985 which was published in the September issue of that year. Two years later Miss Gullanartsiri sent us a New Year 'present' from Thailand in the form of the 1987 listing only to be upstaged by Oliver Anderson of Pennsylvania State University who submitted an alternative *in order!* The topic now enjoys a well deserved rest from the pages of *Mathematics in School* (possibly to be revived as a real challenge in the year 2000?) but it serves to illustrate the extent of teachers' concern with exercises which can involve a wide spectrum of pupils while at the same time making them think and, above all, motivating them to *do* mathematics.

BICYCLE MATHEMATICS

The paper discusses some applications of mathematics to bicycle gearing, speedometers and bicycle braking. A novel manner of calculating bicycle speed is developed.

by William D. Newton, Education Department, Victoria, Australia

Introduction

Throughout the world today there is a bicycle boom. Men and women are out trundling while the ever hyper-energetic youngsters are busy wheeling and skidding their 10-speeders and dragsters.

The bicycle was popular from the turn of the century until the 1920s, but it lapsed into a marked decline with the increasing popularity of the motor-car. Now the wheel has turned, as more and more people see both the personal and broad social benefits accruing from the use of this efficient machine. Riding a bicycle is cheap, it contributes to good health and most importantly, in times of heightened ecological consciousness, it does not pollute.

Teachers might like to capitalise on the enthusiasm of youngsters for the bicycle. As a unifying theme it provides a sound vehicle for excursions into such diverse areas as pollution, fuel conservation, physical fitness, exercise and weight control, ergonomics, recreational activities and the use of leisure time, road safety, bicycle paths and bicycle mechanics. In fact, "bike" could be an interesting subject to introduce into school curricula. The boundaries of the subject would probably be less circumscribed than those of say, history or mathematics.

Bicycle mechanics offers enormous opportunities for teachers of mathematics and science. Some topics could even be of interest to your students! I invite you to try the following ideas with them.

The principle of bicycle gearing

As an exercise to introduce your pupils to the principle of bicycle gearing you might like to set up a practical activity involving the simple machine shown in Figure 1. Jam tins (or press-on-lids) of different sizes and connected by a taut rubber-band provide a good model of the single geared bicycle.

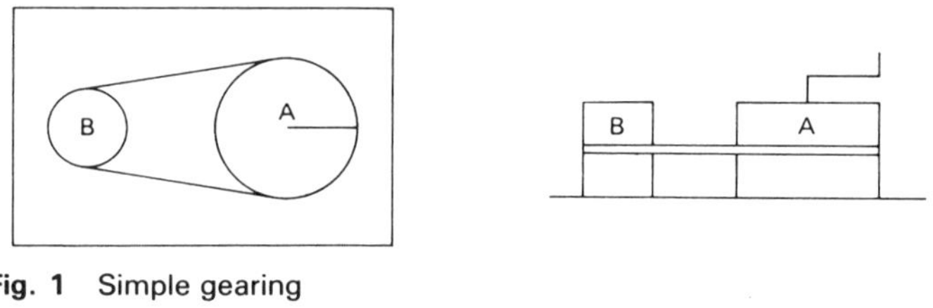

Fig. 1 Simple gearing

Questions to be resolved with this model could include:

(a) Which of A and B (the chainwheel and sprocket respectively) turns more quickly?
(b) How many times does the sprocket turn for 1, 10, 100 turns of the chainwheel?
(c) Obtain the ratio, number of turns of sprocket/number of turns of chainwheel, and compare this with the ratio of the measured circumferences.

From this type of activity pupils can be guided to a bicycle gear ratio formula:

$$\text{gear ratio} = \frac{\text{number of teeth in large sprocket } (T)}{\text{number of teeth in small sprocket } (t)}$$

Gearing of modern day bicycles is usually expressed as either a "gear" in the British system or as a "gear development" in the continental system. Both of these terms relate back to the gear ratio.

The British "gear"

The gear of a bicycle is found using the formula

$$\text{gear } (g) = \text{gear ratio} \times \text{wheel diameter } (d)$$

i.e.

$$g = \frac{T}{t} \times d \tag{1}$$

An appropriate gear for a single geared bicycle might be derived from a front chainwheel size of 49 teeth (i.e. $T=49$), a rear sprocket of 20 teeth (i.e. $t=20$) and a wheel diameter of 27 inches ($d=27$), giving $g=\frac{49}{20}\times 27=66$ (inches).

Historically, the size of the gear in inches, relates back to the "penny-farthing" or "high-wheeler" velocipede. This direct drive machine required the rider to push a large ungeared front wheel of up to 60 inches in diameter. In the example given, a 66 inch geared modern bicycle would be propelled at the same speed as a 66 inch diameter penny-farthing provided that both machines were pedalled at the same rate. That is, the gearing of modern bicycles can be directly related to the diameter of the driving wheel of the old penny-farthings.

The advantage of modern gearing becomes readily apparent when one considers that many modern bicycles have gearing as high as 100 inches. Without gearing, this would be equivalent to pushing a 100 inch diameter driving wheel! What an exercise in dynamic instability!

Of course, the inner leg measurement placed the obvious physical constraint upon the maximum gear attainable with the high wheeler. The slang expression "big wheel" was originally used for those people who could ride a large wheeled penny-farthing!

The continental "gear development"

While the British "gear" can be equated with the concept of *diameter*, the continental "gear development" is equal to the distance travelled with one turn of the pedal cranks. The distance travelled in one turn of the cranks is equal to the *circumference* of an equivalent ungeared driving wheel.

The gear development of the bicycle given in the earlier example can be simply found using the usual formula for circumference of a circle. The circumference of the wheel is equivalent to the "gear development" (g_d) while the "gear" is equivalent to the diameter of an ungeared drive wheel.

i.e.

$$g_d = \pi \times g \tag{2}$$

or, from (1)

$$g_d = \pi \times \frac{T}{t} \times d \tag{3}$$

where d is the diameter of the *geared* bicycle. The gear development of a bicycle with a gear of 66 inches is simply given by $g_d=\pi\times 66=207$ (inches).

Thus a modern-day 66 inch geared bicycle and an "un-modern" 66 inch front wheel diameter penny-farthing would both "develop" the same distance (207 inches) with a single turn of the cranks. Of course, in the continental system, metric units are applied so that typical gear developments for modern bicycles might range from 2.80 metres to 8.00 metres. (For example, a typical maximum gear development could be found using the following sample values.)

$$T=52,\ t=14$$
$$d=0.69 \text{ (metre) (i.e. 27 inch diameter)}$$

From (3)

$$g_d=\pi\times\frac{52}{14}\times 0.69=8.05 \text{ (metres)}$$

A gear development of 8.05 metres is then simply the distance traversed in one turn of the pedal cranks. In this sense, the continental gearing system is easily understood and easily applied. To determine the gear of a bicycle, wheel it along a flat surface while the cranks rotate once. The distance travelled is equal to the gear development. This principle is utilised at bicycle racing tracks. For example, in many events, junior cyclists are not allowed to ride with a gear higher than 81 inches. This gear corresponds to a metric gear development of 6.50 metres ($T=48$, $t=16$). At the conclusion of an event, the winner's bicycle is checked against two lines drawn 6.50 metres apart. If the cranks have not completed one full revolution over that distance, then the gear development must be greater than 6.50 metres. This procedure is certainly easier than counting the number of teeth on chainwheels!

Despite the simplicity of the continental system, Australia, USA and England predominantly use the British gear, even though its origins and interpretations relate back to the now long gone high wheeler. (Someone should survey the origins of the British units of measurement. What a motley but fascinating collection of tales it would be!)

From the point of view of instructing kids, we are better advised to go metric. The interpretation is simpler, and within a year or two, its use will be reasonably widespread. By manipulating formulae (1) and (3) we are also able to relate the British gear (g) to the continental gear development (g_d), viz., $g_d=0.0254\pi\ g$ ($\simeq 0.08\ g$). Stay with the one system though and forget the transformations!

The 10-speed bicycle

The most distinctive characteristic of this popular bike is its externally mounted Derailleur gear system. Mathematical opportunities abound with the Derailleur gear. Let's consider a few that your pupils might find interesting.

Selection of gearing

The 10 gears are obtained from a combination of a double chainwheel and a rear cluster consisting of five sprockets. The number of teeth on the chainwheel and the sprockets determine the 10 gearing ratios. For example, a common combination matches front chainwheels of 40 and 49 to rear sprockets of 14, 17, 20, 24 and 28. The gear development for this combination of sprockets can be found using formula 3. Figure 2 summarises the gear developments for a bicycle with a 27 inch wheel diameter (i.e. 68.6 cm). The bracketed numbers give the equivalent British gear.

	Combination A				
	14	17	20	24	28
49	7.54 m (95)	6.21 m (78)	5.28 m (66)	4.40 m (55)	3.77 m (47)
40	6.16 m (77)	5.07 m (63)	4.31 m (54)	3.59 m (45)	3.08 m (38)

Fig. 2

The high gear development of 7.54 metres would be hard to push but could be managed on wind assisted flat or down-hill riding. The low gear development of 3.08 metre is suitable for hill climbing. Notice the considerable overlap of gears. Effectively this combination reduces to, say, six or seven different gears. A better selection of gears might try to reduce this overlap, and produce a more evenly spaced range of gears. Of course, there are also other factors besides range to consider when selecting suitable gears. Ease of changing gears is critical.

For example, combination B in Figure 3a would require continuous double shifts in order to run through the gears whereas combination C in Figure 3b would require mainly single shifts.

	Combination B				
	14	17	21	26	31
52	8.00	6.59	5.34	4.31	3.62
48	7.39	6.09	4.93	3.98	3.34

Fig. 3(a)

	Combination C				
	14	16	18	20	22
54	8.31	7.27	6.47	5.82	5.29
38	5.85	5.12	4.55	4.09	3.72

(b)

This array of numerical data can be presented graphically, and so highlight the gear changes for each of the three combinations.

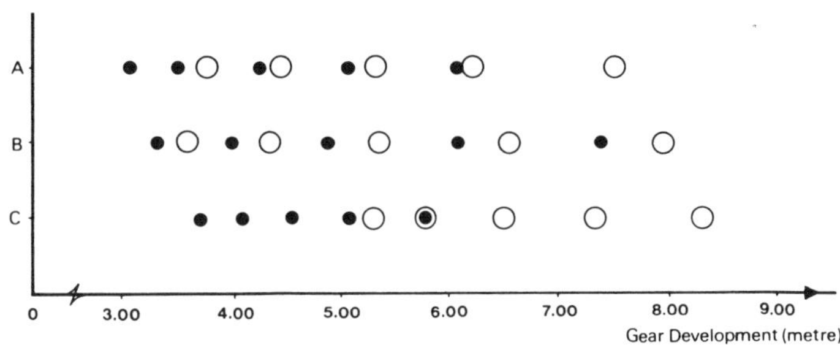

Fig. 4 Graphical representation of gear combinations

The art of riding a geared bicycle well is to maintain an even cadence (pedal rate) regardless of the nature of the terrain. The casual cyclist will probably feel comfortable with a cadence to, say, 60 to 80 pedal revolutions per minute. Intuitively, cyclists realise that their forward speed is related to their cadence. The precise relationship is easily shown and could well provide a meaningful excursion into the use of formulae and their graphical representation.

Consider a cyclist travelling with a pedal cadence of p revolutions per minute. The distance travelled in 1 min will be equal to the cadence, p, by the distance travelled in one revolution of the wheel, g_d. That is, distance equals pg_d. In 1 hour, the distance travelled will be equal to 60 pg_d. This expression is a direct measure of the speed at which the bicycle is travelling in metres per hour. In kilometres per hour, the speed, s, is finally given by:

$$s=0.06\ pg_d \qquad (4)$$

This function in two variables need not be beyond the scope of middle high school pupils. Graphs similar to that shown in

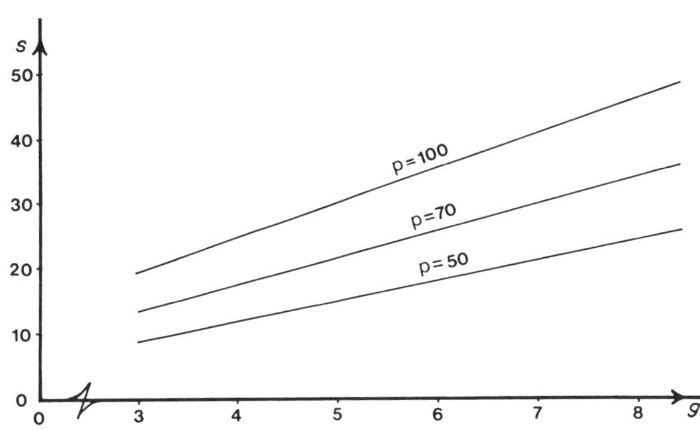

Fig. 5 Determining bicycle speeds

Figure 5 could be developed to show the speeds attained over a range of pedal rates.

A bicycle speedometer

A simple means of calculating bicycle speed can be devised which takes account of both the pedal rate, p, and the gear development, g_d. This speedometer utilises a formula which gives a "time" for each gear such that the number of pedal revolutions in that time gives the bicycle speed in kilometres per hour. Sounds complicated but it's not!

The pedal rate can be found by counting the number of pedal turns (N) in a specified period of time (t seconds)

Then

$$p(\text{revs per min}) = \frac{60N}{t}$$

Substituting this into (4) gives

$$s(\text{km/h}) = \frac{3.6Ng_d}{t}$$

i.e.

$$st = 3.6Ng_d$$

Now, equating the bicycle speed, s kilometres per hour, with N, the number of pedal turns completed in t seconds, gives

$$t = 3.6g_d \quad (5)$$

For example, consider the high gear of 7.54 metres in combination A:

$$t = 3.6 \times 7.54 = 27 \text{ (seconds)}$$

Thus, the number of pedal revolutions completed in 27 sec, while riding in the high gear of 7.54 metres, is equal to the bicycle speed in kilometeres per hour. "Times" for each gear could be similarly calculated and the results taped to the bicycle handlebars for quick reference.

Bicycle braking

Interesting applications of bicycle mechanics are not confined to middle years mathematics. Senior students with an interest in physics or mathematics should find the following discussion on the braking of bicycles a fascinating variation on some familiar topics.

Wheel reactions during travel at constant speed

When the bike is at rest or is travelling at constant speed, both the braking force, F, and the reaction of the cyclist to this retarding force, P, are zero.

Taking moments about the point B gives:

$R_f \times 107 = W \times 43$ where, for example, 90 kg wt is the combined weight of the bicycle and cyclist, W.

This equation gives $R_f = 36.4$ (kg wt), and by subtraction, since $R_f + R_r = 90$, it follows that $R_r = 53.6$ (kg wt).

Fig. 6 Forces on a bicycle during braking

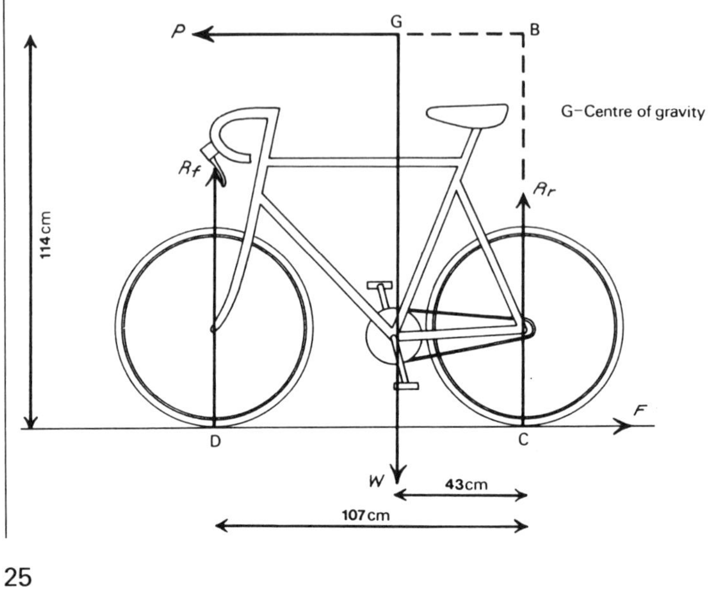

That is the weight of the bicycle and rider is not distributed equally between the two wheels.

Wheel reactions during strong braking

Consider the situation where the bike is being retarded at 0.5 g. This is strong braking, and the experienced cyclist will anticipate the results! By Newton's Second Law, a *total* braking force of 45 g newton or 45 kg wt acts back along the road surface. The reaction of the cyclist on the bike to the retarding force is also 45 kg wt.

By taking moments about C we obtain:

$$R_f \times 107 = (F \times 114) + (90 \times 43)$$

which gives

$$R_f = 84.1 \text{ (kg wt)}$$

and hence

$$R_r = 5.9 \text{ (kg wt)}$$

From this calculation it can be seen that the rear wheel is supporting approximately 6 kg wt of the combined 90 kg wt.

Stronger braking would produce the inevitable unstable situation of the rear wheel leaving the ground! This limiting case is obtained when the direction of the resultant of P and W passes beyond the point A. In this situation, the limiting rate of retardation is $\frac{5}{9}g$. The calculation provides an interesting application of the Triangle Rule of Forces. You might like to try it with some of your senior students.

An interesting example arises with the modern dragster bicycle. You have probably seen youngsters riding merrily around on the back wheels of their bicycles. What construction quirk produces this strange result? Get your students to measure the key dimensions of the dragster and examine the "operating characteristics" of this unusual velocipede. Let's hope we're not too bogged down with prescribed curriculum for students to step outside and apply their mathematics to "live" problems. If time and interest allows, you might further challenge your pupils with such exotics as determining:

(a) The banking required on a circular 400 metre track if the maximum speed of a racing cyclist is about 65 km/h. What would be the *minimum* speed required of the cyclist to ensure that he didn't fall *down* the track.
(b) the terminal speeds reached by cyclists freewheeling down slopes of various grades.
(c) the power output required of a cyclist for hillclimbing.

An enormous variety of bicycle activities remain undiscussed, even within the relatively narrow areas of gearing and braking. For example, who needs a trundle wheel to measure distance when you've got a bike? I hope that this paper can stimulate you to looking at further mathematical and mechanical possibilities of the bike.

Acknowledgement

I wish to acknowledge the valuable advice and assistance given to me in the preparation of this paper by Mr Ron Shepherd of the Curriculum and Research Branch.

Bibliography

Ballantine, R. (1970) *Richards' Bicycle Book*. London, Pan Books.
Metz, J. R. (1975) "Pedalling mathematics." *The Mathematics Teacher*, **68** (6), 495-8.
Whitt, F. R. (1974) *Bicycling Science: Ergonomics and Mechanics*. Cambridge, Massachusetts: MIT Press.
Wilson, S. S. (1973) "Bicycle technology." *Scientific American*, March, 81-91.

MAGIC SQUARES

p	$p+s$	$p+r+s$	$p+2q+r$
$p+q+2r$	$p+q+s$	$2p$	s
$p+q$	r	$p+q+r+s$	$2p+s$
$p+2s$	$2p+q+r$	q	$p+r$

by Mike R. Rose, Ilfracombe School & Community College

Magic squares can prove an endless source of ideas for the mathematics classroom. Even in the era of the calculator and computer their patterns can bring pleasure, simple arithmetical practice and mathematical insight into a lesson. Their use seems to extend from the junior years of the primary school through to university. This article seeks to chronicle some lesson ideas which have been tried out at Ilfracombe School with a variety of age groups and abilities.

Parametric Form

The parametric form of the 3 by 3 magic square is usually found in text books in the form...

$x+y$	$x-y-z$	$x+z$
$x-y+z$	x	$x+y-z$
$x-z$	$x+y+z$	$x-y$

...from which the pattern of the magic number, namely $3x$, can be easily seen. The confirmation of the sum of each row and column is a simple algebraic exercise for year one and two pupils. Most pupils seem to enjoy substituting values for x, y, z into the pattern to produce their own magic squares and one or two year one pupils have even anticipated directed number by choosing $x < y + z$ inadvertently. At the year two or three level the same pattern can be deliberately used by the teacher to develop skills and techniques with directed numbers. At the year four or five level the form can be "re-discovered" by placing the middle x in position and suggesting that the magic number might equal $3x$ and inviting suggestions for a corner value, other than another x of course. The technique of finding the other squares is then not unlike the processes sometimes met in algebraic Venn diagram problems. Finally the above pattern can be used in the sixth form in several quite unusual respects...

(i) as the basis for a simple but pleasing computer programme which prints out magic squares or parts of them for randomly or specifically chosen x, y and z values....

(ii) as a surprising algebraic exercise at the beginning of the sixth form course! If the row products are formed and then added, the result will be found to be the same as the total of the column products. Thus we return to the year one class with yet another idea to try, with or without a calculator of course.

Example

As an example take the case $x=5$, $y=3$ and $z=1$ giving the most basic and fundamental magic square namely

8	1	6
3	5	7
4	9	2

$$8 \times 1 \times 6 = 48$$
$$+$$
$$3 \times 5 \times 7 = 105$$
$$+$$
$$4 \times 9 \times 2 = 72$$
$$=$$
$$\boxed{225}$$

8	1	6
×	×	×
3	5	7
×	×	×
4	9	2
=	=	=

$96+45+84=$ 225

Is the sum of the row or column products always equal to the square of the magic number as in this case?

The above property provides an excellent calculator practice exercise with both larger and negative numbers. The use of the calculator memory can be helpfully encouraged. (The work sheet . . . may be a starter for ideas in this direction.)

4 × 4 Magic Square

To conclude here are two patterns for the parametrisation of four by four magic squares. Unlike the three by three case, which can always be reduced to the algebraic form given earlier, there are many parametrisations of the four by four case which are not isomorphic.

Pattern number one, containing no explicitly negative terms can be used in much the same way in years one and two as the 3 by 3 pattern. The values $p=1, q=3, r=5, s=7$ give the numbers 1, 2, . . ., 15, 16, magic square. Alas the row product and column product property is no longer always true.

Pattern number two gives the so called diabolical magic square of order four by four. P = 1, Q = 3, R = 5 gives the 1, . . ., 16 case.

No. 1

p	$p+s$	$p+r+s$	$p+2q+r$
$p+q+2r$	$p+q+s$	$2p$	s
$p+q$	r	$p+q+r+s$	$2p+s$
$p+2s$	$2p+q+r$	q	$p+r$

⇒

Magic No $=2(2p+q+r+s)$

No 2.

P	Q+R	Q+2R	P+2Q+R
P+Q+2R	2Q+R	2P	−P+Q +R
P+Q	R	2Q+2R	P+Q+R
−P+2Q +2R	2P+Q +R	Q	P+R

Magic No = 2(P+2Q+2R)

Worksheet

In any magic square the sum of the numbers on any ROW is the same as the sum of the numbers in any COLUMN.
The row total or column total is called the MAGIC number for the square.
The sum of the numbers in any DIAGONAL line is also the magic number.

1 a. Copy the following magic square and complete it.

56	83	98
121		

Magic number =

b. Copy and work out these calculations . . .

The product of the 1st row numbers = 56 × 83 × 98 =
The product of the 2nd row numbers = 121 × × =
The product of the 3rd row numbers = × × =

The SUM of the row products =

Hint Add each row product to the memory.
The product of the 1st column nos. = 56 × 121 × =
The product of the 2nd column nos. = 83 × × =
The product of the 3rd column nos. = 98 × × =

The SUM of the column products =

Hint Add each column product to the memory after clearing the total of the row products of course!!!!

2 Repeat the above project for the following magic squares . . .

(a)

56		
168	98	
75		

(b)

− 56		
160	43	
		141

(c)

		1.52
	1.14	0.19
		1.71

3 Make up your own magic square of size 3 by 3. Use a middle number like 100 . . . the MAGIC number is always 3 times the middle no. Try the above project for your square. *State a conclusion for 3 by 3 magic squares.*

DROPPING A BRICK

by R. D. Y. Perrett

"Dad!" said my young son, "the other day when we were out in Derbyshire, my friend and I dropped a stone down an old mine-shaft and it was three seconds before we heard it hit the bottom. How deep would the shaft be?" Little did he know what he was letting me in for. I had done this myself in my younger days, got an approximate answer for myself, and thought the problem to be essentially uncomplicated. I said to my son: "The sound does not take long to come up from any likely depth, and we can more or less ignore it. The formula for gravity-fall is $S=\frac{1}{2}gT^2$. This makes it about 145 feet."

Unfortunately, my infernally active mind would not leave it at that. "Let's try to make some allowance for the time for the sound to come back," I said. "Now at 1,100 feet per second, the sound would take about one-seventh of a second to come up from about 150 feet; so suppose we knock one-seventh of a second off your three seconds, and take it that the stone was actually falling for two and six-sevenths of a second. This (after a little arithmetic) makes it 131 feet." (My son, at 14, is just old enough to have missed the full impact of SI, and like me, he still thinks in feet.)

That was as far as my boy was interested; but some imp drove me to go into the matter further, with some rather surprising results. Let S be the shaft-depth, T the delay-time before the sound is heard, and V the velocity of sound.

First Practical Approximation (ignoring time of sound-travel)

$S=\frac{1}{2}g\,T^2 \ldots$ (A)

Second Practical Approximation

Reduce T by $\frac{1}{V}\times\frac{1}{2}g\,T^2$ and apply gravity-fall formula.

$$S=\tfrac{1}{2}g\left(T-\frac{gT^2}{2V}\right)^2=\tfrac{1}{2}g\,T^2\left(1-\frac{gT}{V}+\frac{g^2T^2}{4V^2}\right)\ldots \text{(B)}$$

Accurate Mathematical Solution

T=Time of fall under gravity plus time for return of sound.

$$=\sqrt{\frac{2S}{g}}+\frac{S}{V}$$

$$T-\frac{S}{V}=\sqrt{\frac{2S}{g}}\ldots \text{(C)}$$

Square this and solve the resulting quadratic equation in S. We get:

$$S=V\left[T+\frac{V}{g}\left\{1\pm\sqrt{1+\frac{2gT}{V}}\right\}\right]\ldots \text{(D)}$$

My first reactions to this were: (i) Why two solutions? (ii) This formula, particularly because the velocity of sound figures in it as a prominent factor, did not seem to bear much relationship with my "practical" formulae (A) and (B).

However, (i) was soon disposed of; when one squared (C) above, one introduced the spurious conditions implied by putting a negative sign on the right-hand side of (C), which would disappear on squaring. Of the alternative signs in (D), it was the positive which was the corresponding spurious solution; so I took the negative sign only.

I assumed, naturally enough, that by expanding the square-root in (D) by the Binomial Theorem — since, for practical shaft-depths, $2gT/V$ would be a fairly small fraction — and taking one, two or three terms, I ought to come up with the equivalent of my successive practical approximations. So I tried; the results were surprising.

First Algebraic Approximation

Put $(1+2gT/V)^{\frac{1}{2}}=1$.

One gets $S=VT\ldots$ (E)

This result implies that one can ignore that part of the time-lapse which is due to the fall under gravity, and assume that it consists essentially of the time taken by the sound to return. This is all right — provided the depth of the shaft is, say, three million metres or more. As this would be about half-way to the centre of the earth, we can regard this as a bit unpractical, to say nothing of the fact that "g" would not be constant down such a deep shaft.

Second Algebraic Approximation

Put $(1+2gT/V)^{\frac{1}{2}}=1+\dfrac{gT}{V}$.

One gets $S=0\ldots$ (F).

So now we have abolished sound as well as gravity. How peculiar! Perhaps the reader can spot the flaws in the first and second algebraic approximations.

Third Algebraic Approximation

If we take the Binomial Expansion to three terms, we at last begin to get into the realms of practicality, arriving at $S=\frac{1}{2}g\,T^2$, the first "Practical" approximation.

Fourth Algebraic Approximation

Going to four terms of the Binomial Expansion gives us:

$S=\frac{1}{2}g\,T^2\,(1-gT/V)\ldots$ (G)

Fifth Algebraic Approximation

Taking five terms:

$S=\frac{1}{2}g\,T^2\,(1-gT/V+5g^2T^2/4V^2)\ldots$ (H).

This differs a little from the second "Practical" approximation (B); by about three-quarters of one per cent. in the "3-seconds" case; but I was still not quite content.

I took the accurate formula (D) and, with the help of my log-tables, calculated the "accurate" depth. It came out at 148.5 feet — obviously less accurate than even my first approximation. It turned out that, because formula (D) with the minus sign is the difference of two relatively large but nearly equal quantities in practice, one needed at least six-figure tables to get a reasonably correct result!

THE NATWEST LOGO

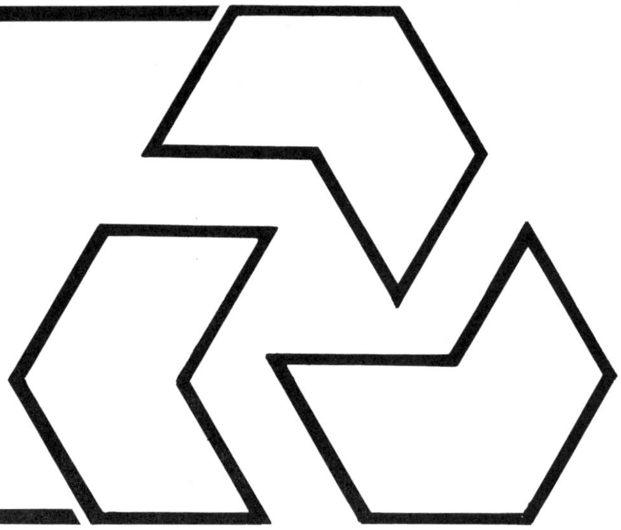

by
Paul Stephenson
Milestone School, London

3.00 p.m.:	**"... and no. 10: an object with rotation symmetry order 3, but no mirror symmetry."**
9.00 a.m.:	*"Sir, I couldn't find anything for no. 10."*

Without having to repeat this homework assignment himself over many years, neither could 'sir'. His collection now comprises a staggering 7 items,[1] and one of the 3 most generally accessible, the Natwest logotype, forms the subject of this article.

What he discovered only recently is that the 2-dimensional motif:  is derived from the isometric projection of a 3-dimensional form:  .[2]

3 Dimensions as 2

There's a 1-many correspondence between the set of 2-dimensional representations and the set of 3-dimensional objects.

Here is a skeleton cube in cabinet projection. Is the marked face at the front or at the back? 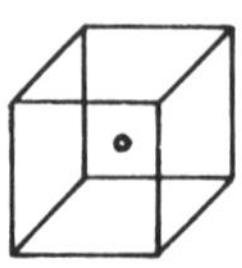 [3]

Here we are looking from the north at a cuboid of which part has been offset along a face diagonal:  [4]

If, in the appropriate diagonal plane, we moved 70° round to the west, would we see ...

this:  ... or this: 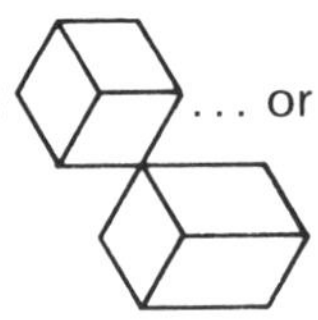 ... or this: 

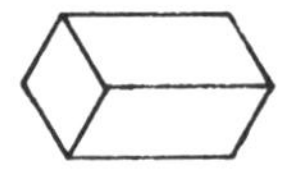

... or ...?

In these two examples I could be said to have made a visual pun: I've shown you 2 – or more – "meanings" for the same "word".

The Natwest chevrons: are so patently planar, it doesn't occur to us they *have* a 3-dimensional "meaning". Hence the pleasure of surprise when one is revealed. The skill of the designers has been to omit these key lines:

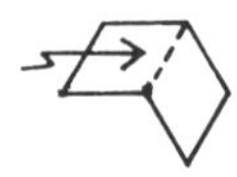

With these inserted, solid realisations – however implausible – might at least suggest themselves: ... the latrine block at a campsite:

... books being hurled (symmetrically!) across the room by a frustrated student:

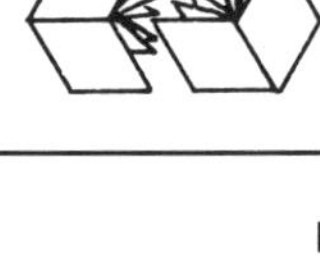

Again, the Natwest designers have been more skillful.

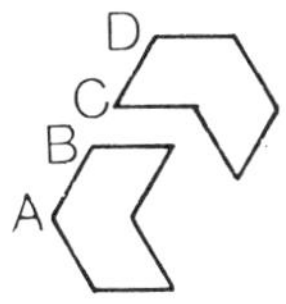

The line AD is illusory. As in my cuboid example, AB and CD are in reality offset.

Working in 3 Dimensions

You should build the cube cluster from interlocking cubes[5] because we're now going to view it from different angles.

Looking at the cluster as it appears in the logo, you see it has the same symmetry as the ring of chevrons, C_3.

Setting up basis vectors in the following (arbitrary) orientation, all of magnitude 1 side:

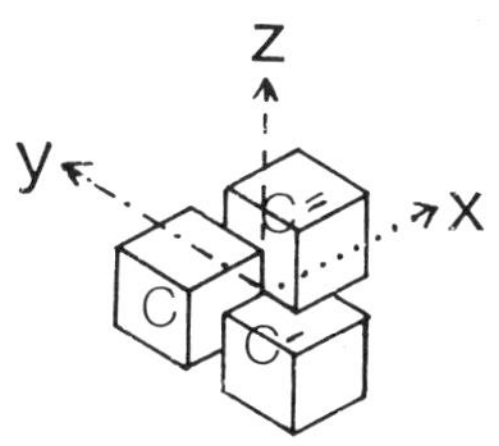

we can form the cluster by taking the cube C

$$\text{and translating it} \begin{pmatrix} \frac{1}{2} \\ -1 \\ -\frac{1}{2} \end{pmatrix} \text{to C}'$$

$$\text{and translating C}' \begin{pmatrix} \frac{1}{2} \\ \frac{1}{2} \\ 1 \end{pmatrix} \text{to C}''$$

You can work out that $\begin{pmatrix} -1 \\ \frac{1}{2} \\ -\frac{1}{2} \end{pmatrix}$ takes C″ back to C, and that all these vectors – as you would expect from the symmetry – are of equal size, namely $\sqrt{1^2+\frac{1}{2}^2+\frac{1}{2}^2}=\sqrt{\frac{3}{2}}$. In other words, corresponding points on the 3 cubes define an equilateral triangle. The particular one formed by the cube centres is symmetrically disposed with respect to the 3-axis of the cluster. Furthermore, if we put the whole thing in a 2^3 box, we preserve the 3-symmetry; that is to say, the triad axis is a space diagonal. All this is clear from isometric projection:

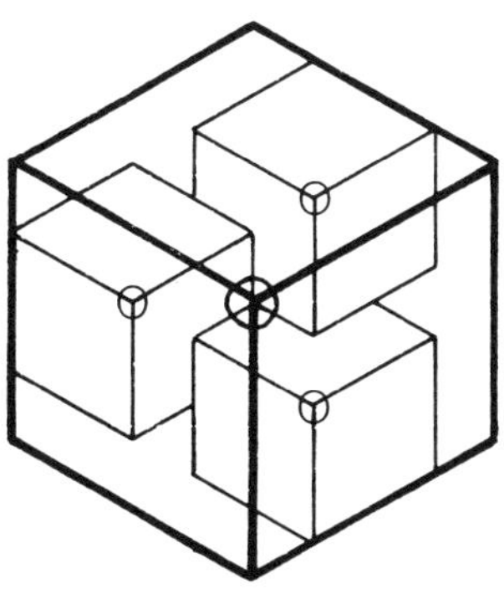

Having stuck 2 cubes together, we can add the 3rd in either of the dotted positions, giving rise to the left- and right-handed forms. In 2 dimensions this corresponds to using the template the wrong way up.[6]

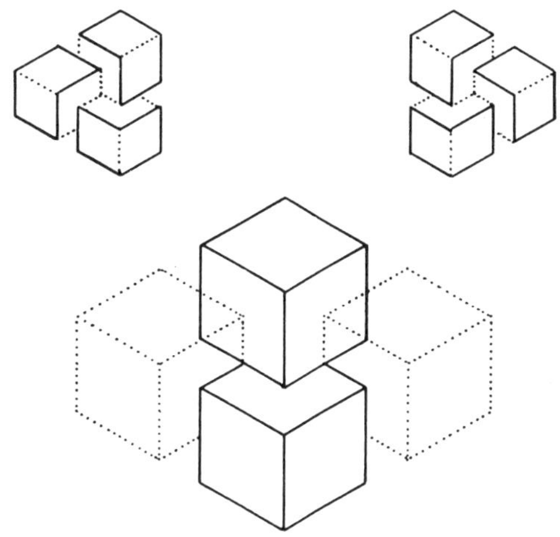

We shall return to this question of handedness.

An interesting exercise is to decompose your model into congruent units wile preserving the 3-symmetry. Such units will be n-cubes for which $n|24$. Any solution for which $n \nmid 8$ would require a shape which itself had 3-symmetry. Unfortunately, the only one which has, this 6-cube:

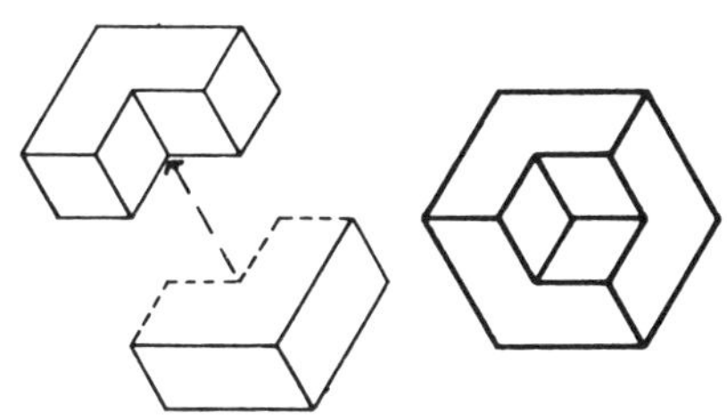

does not replicate. This restricts us to 2-cubes, 4-cubes and 8-cubes. Each of the 3 constituent cubes can be partitioned in this way without difficulty – and without interest. More pleasing is to seek an arrangement in which the units cross cube boundaries.

By dismantling our unsuccessful 6-cube and its 3 satellites into 2-cubes, we can find 6 such arrangements for each mirror form of the cluster:

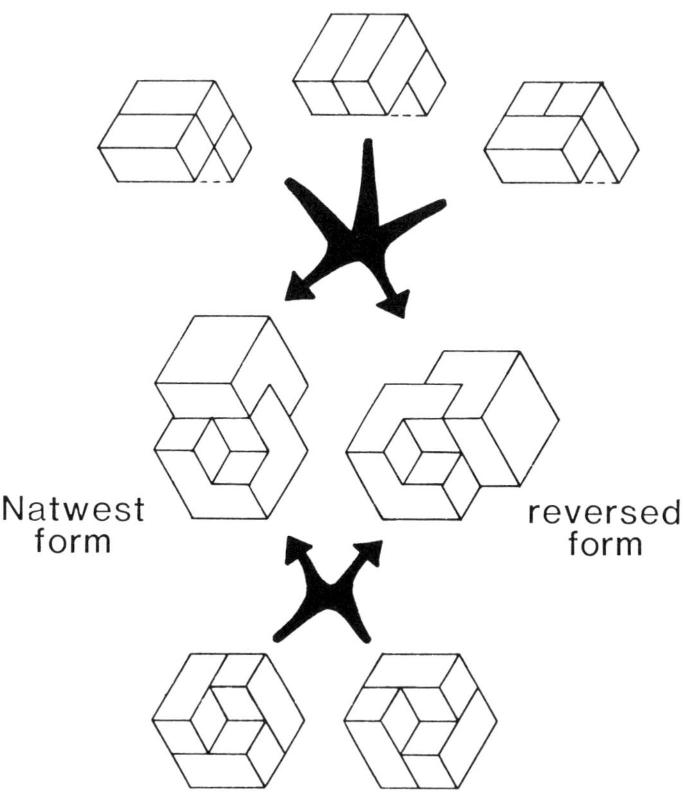

I can't find a 4-cube solution,[7] but by joining 2 non-congruent 4-cubes we arrive at a neat 8-cube dissection. The 8-cube can be made up in 2 ways. Here I show this. I also show how the handedness of the components is expressed in the handedness of the enantiomorphs:

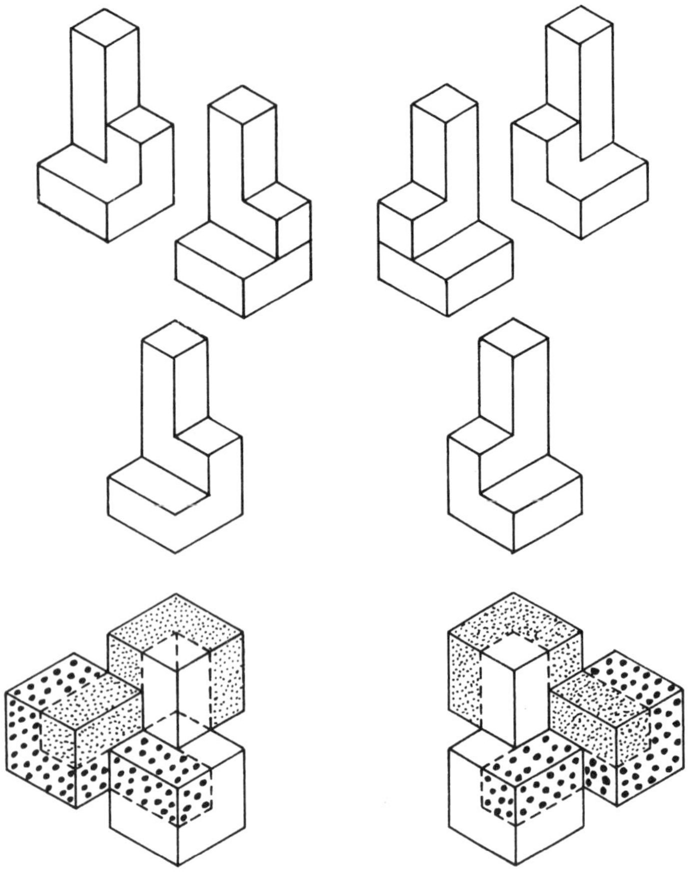

We might enquire whether the cube cluster tessellates. Start with a 6^8 array.

You will see that we can happily translate the stippled layers $\begin{pmatrix}k\\0\\l\end{pmatrix}$, $k, l \in \mathbb{R}$. Having done so, however, unless $k, l \in \mathbb{Z}$, there is no way we can push any cubes in the y-direction: the thing is jammed:

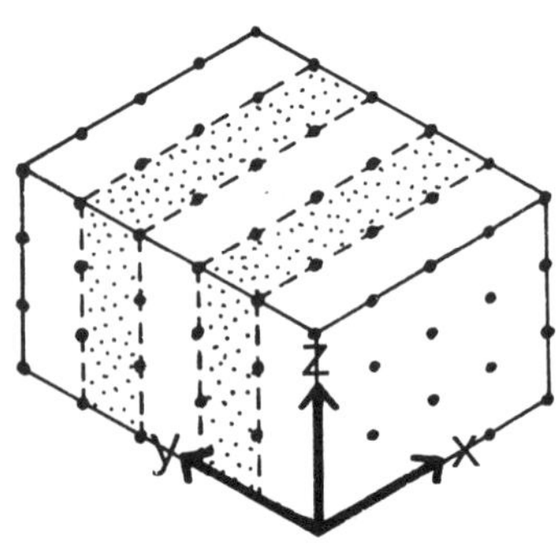

I hope you can find something here to use with your pupils. I would be particularly interested to hear from readers who use the cube cluster as starting-point for an investigation with children who have followed Geoff Giles' courses.[8]

Notes

1. **The "3 most generally accessible" are:**

- the roundabout sign: and the road markings for a mini-roundabout:

  (reversed in countries where they drive on the right.)
- the "pure wool" symbol:
- and the Natwest logo:

More localised – in whatever sense – are:

- the 3-limbed swastika or triquetrum, shown here as the Legs of Man, but taking many other forms. See Hermann Weyl: *Symmetry* (Princeton 1952), p. 66.
- See Martin Gardner: *Further Mathematical Diversions* (Allen & Unwin 1969), pp. 24–5 for a note on this symbol.

The Three Borromean rings

- bank giro credit
- Buckinghamshire College of Higher Education

Readers might use the letters column to push up this modest score.

RECYCLABLE PACKAGING

This looks as if it ought to have 3-symmetry, but a piece of tracing paper superimposed and given a 1/3-turn reveals that it hasn't. The thing is, it's a Möbius strip: follow the arrows round.

2. The cluster symbolises the 3 banks which amalgamated in 1970 to form the National Westminster: the District, the National Provincial and the Westminster.
(The 3-D form is not to be confused with the 2-D form in bas-relief!)
3. Experimental psychologists call it the Necker Cube.
4. For other tantalising examples see Geoff Giles' Build Up and 3-D Sketching Project Packs. A child who has followed these courses would be ideally prepared for my 3-D exercises. The packs are available from: Tarquin Publications, Stradbroke, Diss, Norfolk IP21 5JP at £10.15 and £10.70 respectively, postage £1.50.
5. You'll need to make each Natwest cube of edge 2 units:

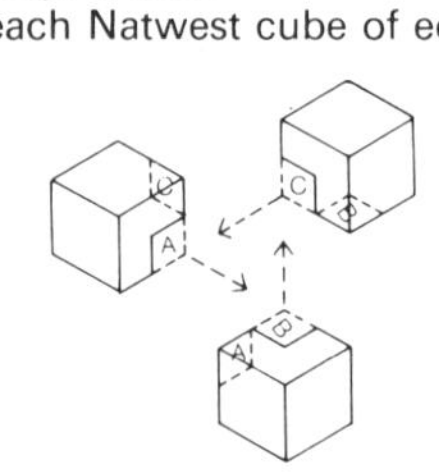

Suitable are:

1 cm: Centicubes (Osmiroid),
Minicubes (ESA);
2 cm: Multilink Cubes (ESA)

6. Mr D. Witchell, the Advertising Manager for Natwest – to whom I am indebted for all my background information – tells me this happens!
7. Perhaps one of your pupils can settle this question – by the method of exhaustion 'or otherwise'.
8. See note 4.

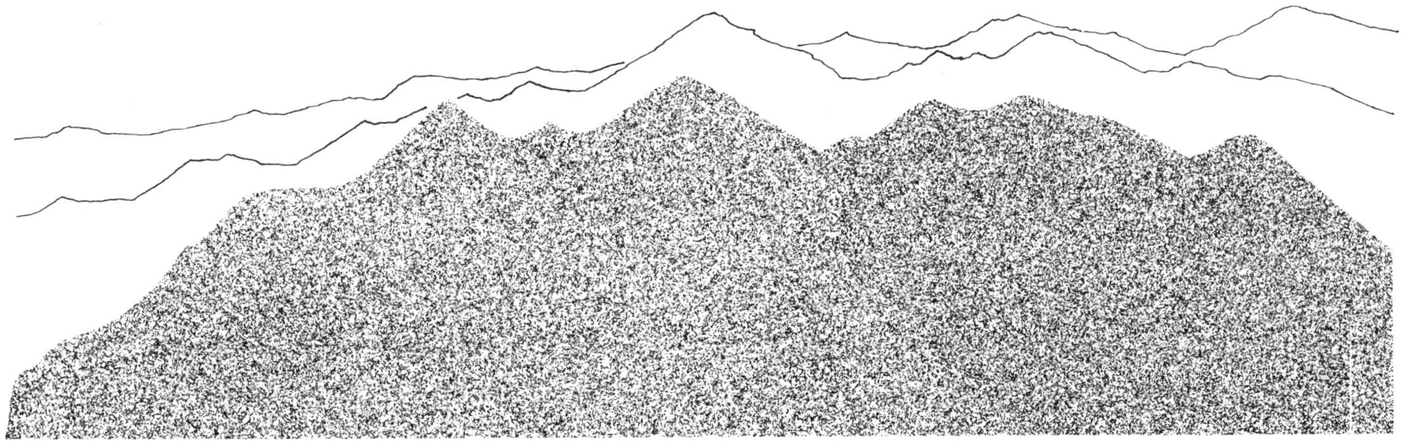

MOUNTAINEERING MATHEMATICS

by Nick Mackinnon, Winchester College

Ridges and Faces

The history of climbing in the Himalayas and the Alps indicates that it is the ridges of mountains that get climbed before the faces. The Matterhorn for example was first ascended by Edward Whymper and his group in 1865. On the descent four of the party were killed when a rope broke. The North face wasn't climbed until 1931 when the Schmid brothers just managed it.

This can be explained with some three dimensional trigonometry. In the pyramid below the angles of climb up the two faces are 76° and 73°, while the angle up the ridge is 69°.

A problem that arises now is what function of the angle of climb gives the best measure of comparable difficulty?

Fig. 1

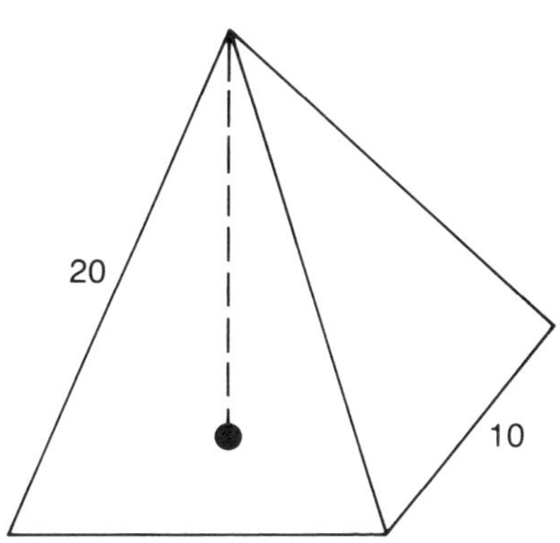

Traversing a Snow Slope

A related example is the way a mountaineer will traverse across a snow slope. Here there is a play-off between the decreasing angle of climb and the increasing distance that has to be walked. This becomes critical on a climb like the North Face of the Eiger where the stonefall by day makes a slow climb too dangerous. Bonnington's account in "I Chose to Climb" describes how he and Clough cut up the second icefield while a second group who traversed were swept to their deaths.

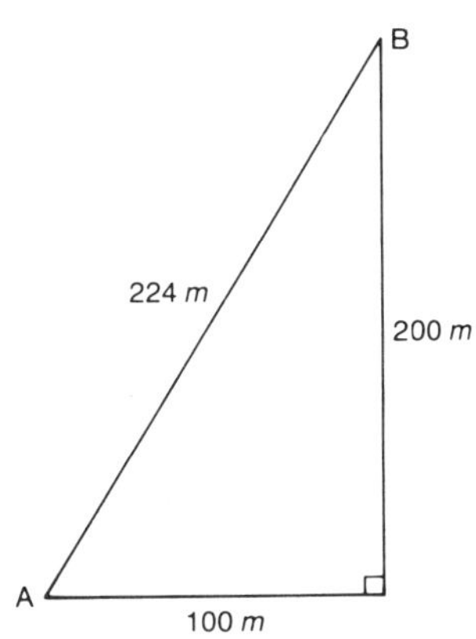

Direct route AB. 224 *m* at angle 63°.

Fig. 2

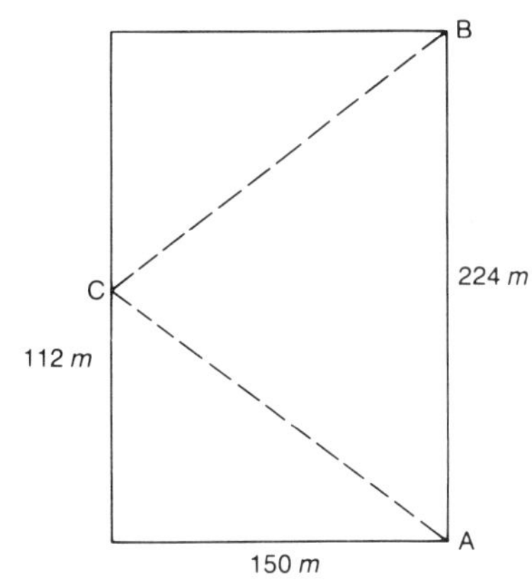

Traverse route ACB 374 *m* at angle 32°.

Fig. 3

Schichallion

Newton suggested that the deviation of a plumb line by a mountain's mass could be used to determine the constant of universal gravitation, and hence establish the mass of the Earth. After much surveying the mountain Schichallion on the South East of Rannoch Moor was selected because of its isolation and symmetry. In 1774 the Royal Society backed Maskelyne, the Astronomer Royal, in his survey. It took two years and the sighting of 43 stars through the 10-foot telescope which took a gang of 12 men to shift around the mountain. One of the surveyors, Hutton, invented contours during the mapping of the mountain.

It appears on sheet 42 (Loch Rannoch) of the 1:50 000 O.S. maps, and calculating its volume is a nice exercise. On my next trip to it I plan to take some rock samples and calculate the mass of the mountain. The mountain is an easy walk from the car park in the Braes of Foss (where a plaque commemorates the experiment) and would be a good excuse for a field trip ...

Avalanches

The most dangerous slopes for avalanches are those between 30° and 45°. (Why?). The prediction of avalanches is helped by the use of a penetrometer which measures the resistance of the snow. More details of this technique appear in Langmuir.

The search for victims is still best done by dogs. Failing this, probing of the snow is done. It is an interesting Operations Research problem to decide the fineness of the probing. There are two main methods in use. The searchers stand with their feet 50 cm apart and separated from their neighbours feet by 25 cm. For a coarse probe each searcher probes once between his feet and then walks 70 cm forward. With 20 searchers one hectare can be searched in four hours and the success rate is 76%. In a fine probe the rods are driven in at both toes and also between the feet. The line them moves forwards by 30 cm. It would take 20 men 20 hours to search one hectare and the success rate is 100%. The choice will depend on the assumed depth of burial and the condition of the snow (whether for example air pockets are possible).

The diagram below (Fig. 4, adapted from Langmuir) shows the probability of survival. It would be an interesting project to investigate various probe procedures and their probabilities of success. As a matter of interest the best strategy if caught in an avalanche is not to struggle until the last few seconds when an attempt should be made by swimming movements to minimise the depth of burial.

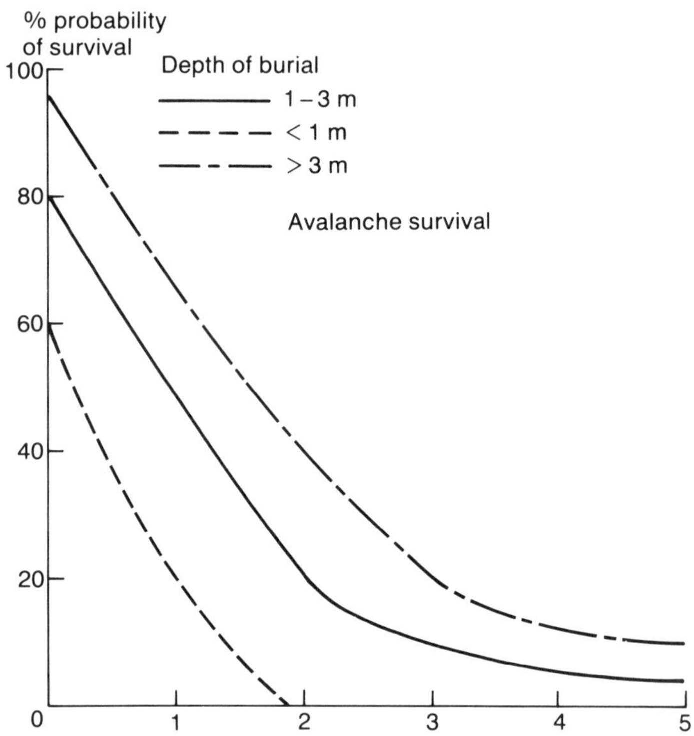

Fig. 4

Searching in Mist

If the limit of visibility is one grid unit the spiral search procedure shown below (Fig. 5) is efficient. Notice that every point lies within one grid length of the path. Is it the most efficient? This will depend on the probability distribution for the location of the objective.

Naismith's Rule

Naismith's rule: on mountain territory allow 15 minutes per kilometer plus one minute for every 10 m of ascent.

One nice use of the rule is to give groups an expedition to plan (for example the summits of the 14 Welsh 3 000 peaks, O.S. 1:50 000 sheet 115 Snowdon) with the challenge of minimising the time. The record for the Welsh 3 000's is an astonishing nine hours (starting on the summit of Snowdon). Of course a much speeded-up Naismith's rule is needed by these supermen.

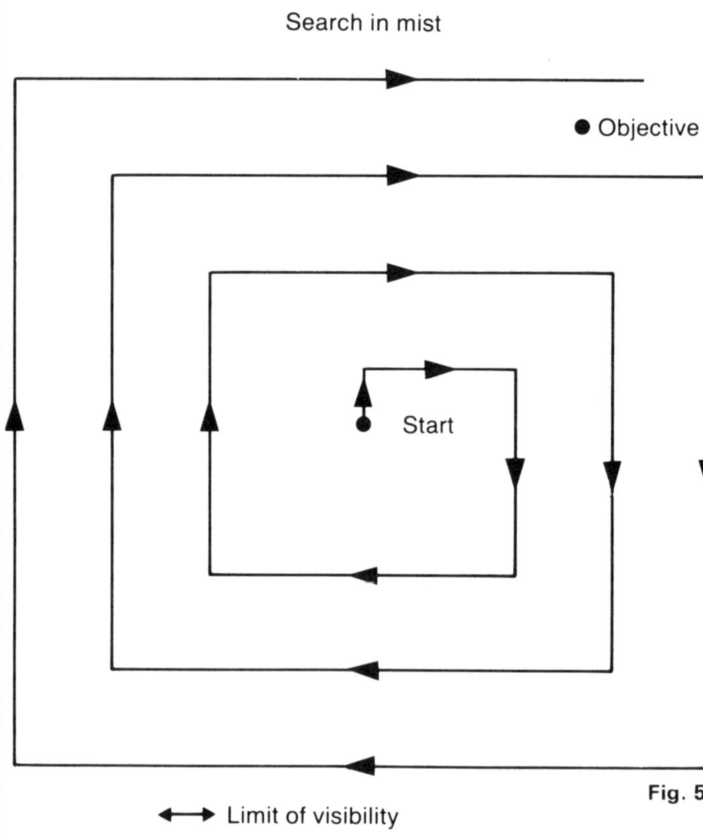

Fig. 5

Chill Factor

A very light wind has a surprising effect on cooling. Heat transfer in air is mainly by convection not conduction. If there is no wind, temperatures of −20°C are found to be only "cool".

The nonogram below is reproduced from Langmuir.

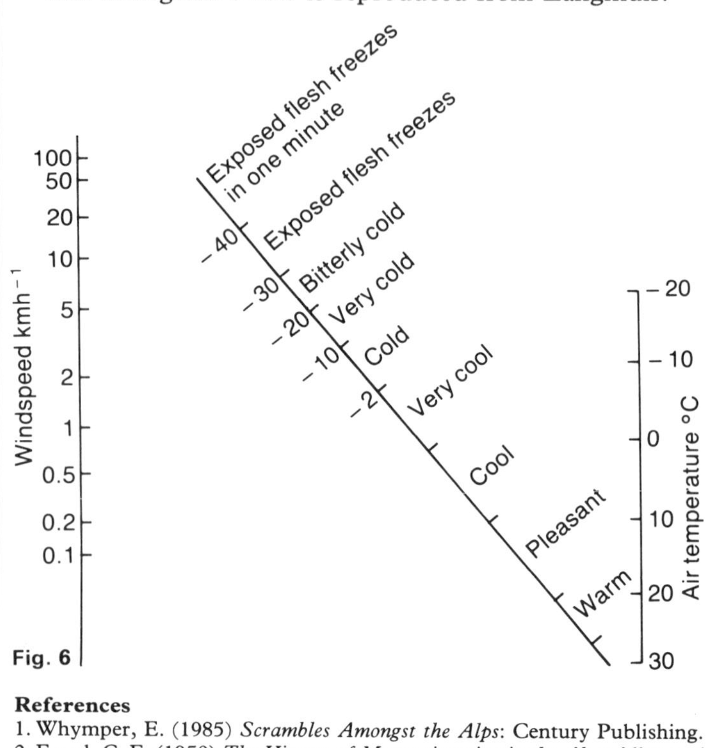

Fig. 6

References

1. Whymper, E. (1985) *Scrambles Amongst the Alps*: Century Publishing.
2. Engel, C. E. (1950) *The History of Mountaineering in the Alps*: Allen and Unwin.
3. Bonnington, C. (1985) *I Chose to Climb*: Gollancz.
4. Langmuir, E. (1984) *Mountaincraft and Leadership*: The Scottish Sports Council and Mountain Leader Training Board. (This book contains much else of mathematical interest.)

MATHEMATICS INTERNATIONAL

by J. A. Glenn

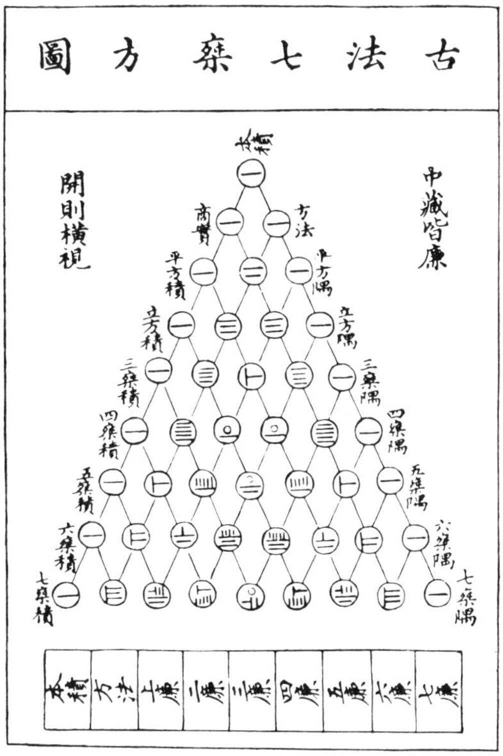

A Suggestion for the Classroom

The notation of mathematics, like that of music, is more or less independent of the language of its context. An expression such as $x^2 + 5x + 6$ is read aloud by us as English but is equally intelligible, not only in Germany or France, but in Japan or Israel whose languages are not written in the Roman alphabet.

Starting from this point we can make up some exercises which the more able children should find interesting, exercises which can not only help convince them that mathematics is truly international, but gives them an approach to foreign tongues uncommon in the lessons usually devoted to them. This approach featured in the lively BBC "Russian Language and People" television series, one of whose aims was to "convince the student that he can extract information from a Russian text".

The suggestion is that many mathematical texts contain enough clues to allow one, if not actually to translate, at least to get a very good idea of their meaning, even if without prior knowledge of the language in which they are written. This is surely a valuable skill worth practising in its own right. At another level, given an elementary knowledge of the language, a simple problem in arithmetic gives sentences not constructed for the benefit of learners, but given in a meaningful context expressly for native speakers. Where, other than in a French schoolbook, would you find written for French eyes such a sentence as "Une fillette a trois pommes."?

The project is easy enough, given co-operative friends who like bookshops and can afford foreign travel. Many of the shops will stock elementary textbooks, whose level can be judged from the format and the illustrations or diagrams, if not from the text. Any one of these books can suggest work that bright pupils seem to find it fun to puzzle over. What the exercises test can be argued about, but surely one cannot deny the value of being able to impose sense on what is at first glance not intelligible.

A few examples of workcards follow. Any cards in French or German, for pupils who are studying these languages, can obviously be expanded into exercises that are "real life" in the sense that French or German children have to cope with them. An example has been taken from Norwegian because it is cognate with English, and from Japanese because without the mathematics it is likely to be impenetrable. Avoid the word "translate" in making up cards. A child might be expected to guess that *skolebanken* could be rendered as "school bank" but only grammatical knowledge would arrive at the correct version "the school bank". No language need be excluded if the mathematics of the passage chosen is sufficient to suggest meaning.

Answer each of these questions in French. Your answer should be a complete sentence.

1. *Papa avait 67 noisettes. Il les partage entre Paul et Louis. Il en donne 34 à Paul. Combien en donne-t-il à Louis?*

2. *Le papa a partagé un sac de noisettes entre Pierre et Jean. Il en a donné 48 à Pierre et 49 à Jean. Combien le sac contenait-il de noisettes?*

In Norway money is reckoned in kroner and øre, with $100øre = 1kr$. Try to guess the meaning of this problem set in a Norwegian arithmetic book. To start you off, Ragna is a girl's name, *fikk* means "got", *mor* means "mother" and *penger* means "money". You should get the rest if you think about it.

Ragna fikk 3kr 16øre av mor og 4kr av far. Hun satte 6kr 10øre i skolebanken. Hvor mange penger hadde hun igjen?

Now try this one. The Norwegian word *kjøpte* means "bought".

En mann kjøpte en traktor for 4875kr og en moped for 1960kr. Hva betalte han til sammen?

Write them both out as if they were problems in English. You can use different sums of English money if you like.

Here is a problem from a Japanese schoolbook. Look at the symbols of number and measure in the text and compare them with the diagram. Write down what you think is labelled by A and what the 15 m probably gives us.

Try to write the whole problem out in English and then solve it.

長さ180mの電車が秒速15mで走るとき，長さ210mの鉄橋をわたり終わるのに，何秒かかるでしょうか。

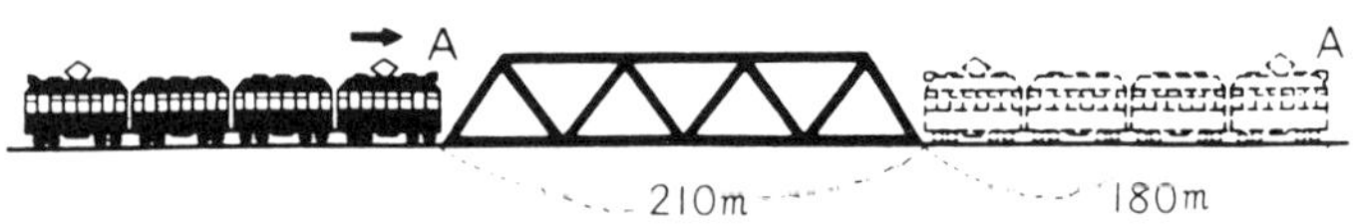

電車の先頭Aが鉄橋をわたりはじめてから，電車全体が鉄橋をわたり終わるまでに走る長さをもとにして考えましょう。

Here is a section of a map and a question in Japanese based on it. Write down what you think the group of symbols on the left stands for.

Go through the text and underline this group of symbols in the three places where it occurs. Then underline the other groups of symbols on the map where they occur, and write down what they probably stand for. Make a guess at the problem the children are asked to solve and write it out in English.

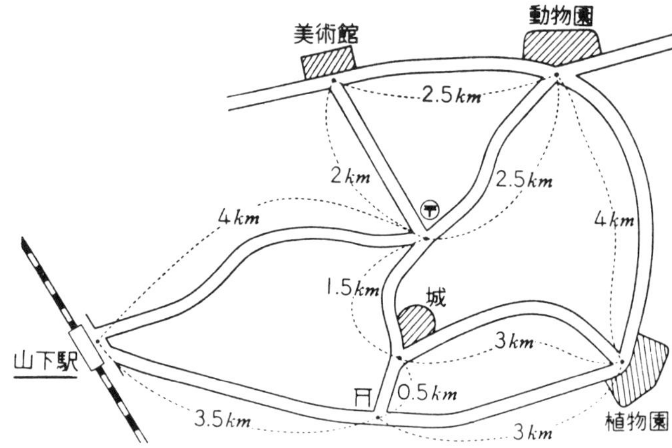

① 上の図は，山下駅に立ててある案内図です。

よしお君は，おとうさんと自動車で山下駅から美術館，城，植物園，動物園をまわって山下駅に帰ってくることにしました。

A View of Mathematics from the Fifth Form

by Giles Dickins and Michael Wood,
University of Oxford, Department of Educational Studies

Introduction

During the spring term of 1981, as part of an SSRC-funded project on pupils' choice of mathematics[1,2], we interviewed 55 fifth-form pupils in five mixed comprehensive schools in Oxfordshire about their views on the subject and how it is taught. We attempted to pick pupils who were either planning to do 'A'-level Mathematics, or who looked as though they might have been able to but were *not* doing so.

We hope teachers will not take our approach as indicating that we accept at its face value all that we were told. Although our main impression is of the pupils' considerable maturity and sensitivity, there are obviously areas where "teacher knows best"; we are not advocating ultra-democracy in the running of mathematics departments. We do believe, though, that the feelings and opinions of the pupils, whether sensible or not, are important information for anyone teaching them. We hope that, by acting as sympathetic listeners, unconnected with the school, we may have helped to uncover more about their views on mathematics.

Mathematics and Other Subjects

Many of the pupils interviewed gave, as a reason for liking maths, the fact that maths involves "working things out". This was sometimes contrasted with subjects such as history, English literature, geography, French and biology which tended to draw comments (from some pupils) like "notes and learning things" and "too much learning" and "tedious". Maths, on the other hand, was "easier to remember". Related to this is the idea — which turned up frequently — that maths, to a greater degree than most other subjects, "explains itself" and "is more logical".

The clarity with which the correct answer is generally defined in school mathematics appealed to some pupils. One girl contrasted mathematics, where you "finish at the answer" with biology which goes "on and on". Another girl contrasted mathematics favourably with history where answers are not right or wrong in the definite sense that they are in mathematics. Someone else, however, put the opposite point of view:

> "*There's no discussion in maths — at all ... There's only really one straight line through things: in maths it's either right or it's wrong ... and you can't have any other ideas about it, whereas in history a lot of it can be what you believe in ...*"

Another pupil described the stereotype of a maths lesson,

which in fact seemed far less prevalent than might be expected, as involving "nothing practical", merely "sums from the board". On the other hand this girl did prefer maths to German, French and English because these subjects involved "learning", presumably in the sense of rote memorization, and, by implication, maths did not. Someone else described doing maths in the same terms as one might use to describe watching a film:

> *"I like to be able to sit down and get into a long sum."*

Many specific topics were mentioned favourably and unfavourably — often the same topic was mentioned favourably and unfavourably by different pupils. Algebra drew some interesting reactions — it was mentioned favourably by six pupils and unfavourably by another six. Some favourable comments suggested the pleasure of being able to do something which looks impossible:

> *"(Algebra) looks complicated but it's not really ... makes you feel intelligent because you can do it."*

Most of the pupils interviewed thought that the maths they learned should be useful in some sense (although there was a tendency to take it on trust that the topics they were learning were useful). Some surprising examples of topics which appeared to be useless were mentioned — probability, standard deviations, matrices and calculus. The last of these in particular appeared irrelevant to some pupils. This suggests that it would perhaps be a good idea to think more about the task of drawing attention to the practical uses of areas of mathematics such as calculus — which may be difficult in the initial stages of teaching the topic.

Making Maths Lessons Interesting

We asked the pupils for their views on how maths can be made more interesting. The importance of relating mathematics to something practical or interesting, or to something with which the pupils were familiar, came through strongly in many of the comments.

> *"If you're given something you have to work it out about something practical that's quite interesting ..."*

Someone else, talking about matrices, commented that she couldn't "see any practical use ... What's the point? ... just meaningless numbers ... have to know what numbers are." Another suggestion, which may not have occurred to many pupils (and might not be taken seriously at least until they had experienced it) was:

> *"Trying to involve pupils as much as possible — say, try to get a particular child to teach for a while just to see how it goes — it'd be different ... just something I've thought of."*

The open question was supplemented by a list of suggestions about which pupils were asked to comment favourably or unfavourably (omitting any about which they had no particularly strong feelings). The following suggestions received many more favourable mentions than unfavourable ones.

1. Feeling you are making progress (27 for, 0 against)
2. Seeing how maths is used (26, 1)
3. Not spending too long on one topic (25, 6)
4. Having problems to solve (18, 1)
5. Mastering difficult ideas (14, 0)
6. Carrying out investigations (12, 1).

Clearly the meaning of these in practice would vary from pupil to pupil. In particular, different pupils obviously had different types of problems in mind. Several did not want problems which were too hard or which involved non-standard methods of solution. One girl said that her "problem" in maths was dealing with problems set in a "situation" (e.g. "man walking") because she did not know what method to use, whereas she could deal with a problem which "says clearly what it's at." This type of difficulty, which is surely very common, conflicts with the requirement to make maths more practical and to apply it to real situations — but this is surely an argument for devoting more time and energy to applications.

On the other hand, as always, the opposite point of view was brought up by some pupils. One enjoyed applying mathematics to her father's business problems, and one boy put in a plea for difficult, complex problems with "lots of stages". He did not like the sheets of short, similar exercises which he was sometimes given. (He had already obtained a grade "A" in 'O'-level mathematics.) The difference between this boy, and those who prefer easy problems, illustrates a very important distinction between being interested in what you feel confident of being able to do without too much difficulty, and being interested in a challenge — in what you feel you may not be able to do at all, or only with considerable difficulty. The "challenge" mentality is obviously worth fostering but we should remember that one pupil's challenge may be another's discouragement.

Some of the other suggestions for making maths interesting elicited more mixed feelings from the pupils. "Using a computer in maths" had thirteen favourable mentions and thirteen unfavourable. The reasons behind the negative comments were the feeling that computing was "a bit terrifying" or only for experts, and the feeling that computing wasn't really maths, or that it was a means of avoiding the real business of mathematics — which for some pupils appeared to be working through long sequences of routine manipulations. This serious attitude towards maths was also apparent in the comparatively large number of unfavourable responses to the suggestion of having "puzzles and games" (sixteen favourable, eleven unfavourable) and to that of the "teacher using models and films, etc." (ten favourable, thirteen unfavourable). The feeling tended to be that these activities were appropriate for younger children and lower abilities, and perhaps occasionally at the end of a term for the pupils being interviewed, but were not appropriate in a serious maths lesson. On the other hand many of those interviewed said they did enjoy puzzles and games in their own time, so it seems a pity that this source of enjoyment is not tapped more in formal maths lessons.

Understanding Maths Lessons

There is a clear link between interest and understanding. In one pupil's words: *"I think that if it's more interesting it's more understandable."* — and, presumably, if you can understand it better it becomes more interesting.

As before, we asked the pupils how they thought they could best be helped to understand the maths they were taught. The importance of practical applications and of relating mathematical topics to everyday things with which the class would be familiar was mentioned by several pupils in the context of understanding as well as that of interest. In a similar vein new topics should be introduced gradually and related to what the pupils knew already. Several pupils mentioned books — in some cases as a supplement to the teacher's explanation, in which case it may be more helpful if the teacher does not simply follow the text but provides an independent approach — but one boy at least found reading books more helpful than learning in class from the teacher. (He obtained a grade "A" at 'O' level.) Perhaps pupils should be given more encouragement to read books about mathematics?

Again, as for the interest question, we gave each pupil a

list of suggestions for improving pupils' understanding of mathematics, and asked for comments. The most popular suggestions were:

1. Being able to work at your own pace (32 in favour, 3 against)
2. Being able to talk to your neighbour about problems (27, 4).

Several people expanded on the second of these:

> *"Pupils can explain something to another pupil better than teacher can."*
> *"If I explain to her (a friend) what I can't do, I suddenly realise I can do it."*

This leads us to wonder whether our pupils have sufficient opportunity to consolidate their understanding of mathematics by talking about it.

Other popular suggestions were:

3. Having enough practice before going on to the next thing (23, 1)
4. Teacher doing more than one example on the board (22, 1)
5. Teacher using diagrams or models (12, 1)

"Having to find out something for yourself" was considered a good idea by those planning to do maths 'A' level (9, 0) but, as might be expected, those not planning to do maths 'A' level were not so enthusiastic (4, 4). Most liked the idea of explaining a method in different ways, although the suggestion of *"Teacher doing a problem in more than one way"* got a mixed reception (nineteen in favour, 13 against). The feeling among those who were against this suggestion was that more than one method for a given problem or type of problem would just cause confusion. The final suggestion — being given apparatus or models to work with — was voted down with four in favour and eight against. This may be a reflection of the same serious attitude mentioned in the "interest" section ("too many people fooling around with them") or a feeling that they were now capable of dealing with abstract ideas and did not need concrete embodiments as younger children, or perhaps their less able contemporaries, might.

Explanations

Many times we were told, in effect, that the teacher's job was to "explain"; we attempted to go beyond this, and explore the pupils' attitude to "relational" and "instrumental" explanations[3]. Clearly we had to avoid using these terms, and so far as possible make the distinction without giving hints as to our own preference. We described a relational explanation as one which covers both what to do and why it works, and an instrumental explanation as one which shows how to get the right answer without going into the reasons. Most pupils appeared to understand the distinction. Those favouring the instrumental approach appeared to take a rather narrow "instrumental" view of mathematics:

> *"That's the object of it isn't it — to get the right answer?"*

Some of those who favoured the relational approach were emphatic about the importance of relational understanding:

> *"If I can do something it's usually because I've been told how it works ... if I'm not told how it works it'll confuse me, it'll really muddle me up."*
> *"Yeah, that's what I keep saying to our teacher — writes up a load of stuff — I don't understand where she got it from and I used to say 'Why does it work?' She goes 'Well you just have to believe me' — stuff like that ... see I can't work like that cos if summat happens I — you know — there's got to be a reason for it happening."*
> *"When I do maths I like to know why it happens — so I can apply what I know (in) that question to something else."*

Most thought, or assumed, that it was the teacher's job to explain why a rule works, but one boy at least was quite happy to work out the "Why?" for himself:

> *"I started off with nothing and derived my own formula (at home) which was the actual cosine formula in the end ... which I liked ... it gave me a much clearer understanding of the cosine rule. He (the teacher) said we didn't need to derive it ourselves but I think even if you don't need to, it gives you a much clearer understanding of the subject."*

This strong insistence on relational understanding, on making sense of mathematics, brings us back to some of the attitudes expressed earlier, that mathematics is liked because it "explains itself" and because in maths you are told "why". In other words mathematics is preferred to various other subjects by these pupils because they can understand it relationally.

The Teacher

The importance of the teacher in affecting attitude and progress was often mentioned by the pupils spontaneously; it was thus natural to turn to this crucial but delicate area, which we did by asking if they had ever had a mathematics teacher whom they considered particularly good or bad. Where several pupils mentioned the same teacher, they tended to agree on whether this teacher was "good" or "bad". There were interesting exceptions, though, where pupils referred to a teacher as "good for me", although the others in the class were said to disagree. We had more favourable than unfavourable mentions, some recalled from primary school.

We then tried to find out why those teachers stood out in the pupil's mind. One of the commonest points mentioned was willingness to explain again, perhaps individually, when a pupil was stuck, and the patience which made the pupil feel it possible to ask for such help; a comment of this kind was made on about half the good teachers, for example:

> *"He was very patient, and if you did ask while you were doing an exercise he would stay with you until you did understand it — he didn't lose his temper at all, and he'd try teaching it from all sorts of angles to try and make you understand."*
> *"If you didn't understand it she would sit down and go over it and over it and over it until you did."*
> *"I suppose the best thing would be to see each pupil in turn and if the pupil understands it well fair enough — let them carry on, but if you get some people who don't it's best probably to teach just those few on that particular subject."*

This contrasts with:

> *"(The teacher says) 'Never mind, just don't answer questions on that in the exam' — (I) can't take that — doesn't make me feel very confident ... I like to sort it out."*

"Feeling you could ask" was important, but some pupils thought they were more inhibited by worry about other pupils thinking them stupid than by fear of the teacher's reaction. A number of teachers were praised for their willingness to offer help outside lesson time.

Some pupils in one school (where tests were frequent)

stressed the importance of tests as another means of diagnosing what individuals had understood, so that appropriate remedial action could be taken. Too much reliance on teaching the whole class together was unpopular with the quicker pupils, too; some of them referred unfavourably to teachers who "went on and on", with the whole class. These pupils also value sensitivity to individual needs, praising teachers who gave them extra "stretching" work — not more drill "to keep them quiet" — when they had finished. Some confessed to a natural human laziness in not actively seeking such extra work, so one teacher, for example, was commended because "he noticed if you sat back", and then "would give you something different to do."

The teacher's expectations were another crucial area. Repeated praise of "strict" teachers seemed to refer more to insistence on academic standards than on disciplinary ones. Thus a "strict" teacher who always marked pupils' books and "made you work hard" was contrasted favourably with one who:

> *"Wasn't very good — he didn't mark our books — he didn't show any interest in our work — we didn't used to work — you know we didn't think he cared about how we did."*

Although some teachers were praised for their tight control and organization, a number of the fifth formers seemed ready for a rather different style. They appreciated the way their teachers were now prepared to give greater responsibility to pupils for organizing their own work schedule. They were quite clear about the consequences which would follow if they misused their relative freedom, and seemed to need little in the way of teacher's reinforcement of this point. For pupils going into the sixth form this style of teaching would seem to be a valuable preparation, and some pupils who were leaving school for college gave as a reason the desire for such increased responsibility. Perhaps we do not always sufficiently adjust our teaching style to the increasing maturity of our pupils?

Many pupils mentioned the importance of the personal qualities of their teacher. A list of all the virtues would be difficult to act on, but it is perhaps worth picking out the frequently mentioned patience, ("doesn't shout"), humour ("you could have a good laugh") and friendliness ("could mix with the kids"); where the teacher's style was more formal than this last quotation suggests, pupils seemed generally well able to detect friendliness and concern underlying the "strict" exterior, or were very disappointed if they could not.

Slightly different from the everyday "sense of humour" was the quality that earned one teacher the comment that he "enjoys it himself". Enthusiasm seems often to be infectious, and so, perhaps, is the lack of it. Pupils praised teachers who gave them some idea of why they were doing a topic, and who gave their own approach to it, which seems to caution against a simple reliance on syllabus and textbook to answer the "why?" and the "how?". For some pupils an important feature of this original treatment seemed to be its visual presentation; thus: "if he wasn't talking you'd be able to see what he'd done each time, quite easily."

If the pupils seem to be asking for perfection, it may be reassuring to find it said of a teacher who "makes slight mistakes" that he also "knows what he's talking about", "keeps us on our toes", and that his mistakes do not shake his pupils' confidence. Confidence, perhaps, is as infectious as enthusiasm, but needs to be distinguished from an appearance of infallibility or of finding everything easy which might be very discouraging for pupils.

Conclusion

We can not claim to establish general truths from work with such an unrepresentative selection of pupils. Nevertheless we feel that these fifth formers may well have expressed views which others would share. We hope that, even though we have found little that is simple and nothing that is new, we may have helped to illuminate this area for other teachers. Certainly we have found the experience of listening to the pupils very thought provoking, and at times very encouraging; for example, this final quotation we find irresistible:

> *"Well, I like the feel of numbers — I like to think that a number actually represents something — I like the idea of that — and, you know, sort of create something inside my head that makes me want to look at it in greater detail ... there's something sort of physical in the way that you can mess around with them and come up with something that actually represents another thing — it's — it's just the way that you can work numbers together to create something, almost."*

The project is still following those pupils in the sample who continue in education. For some of these though, and for all the remainder, their view of mathematics is now, in all likelihood, set for life, as they have probably had their last formal lesson in it.

References

1. John K. Backhouse *et al.* (1982), *Choice of Mathematics for A level*, Dept. of Educational Studies, 15 Norham Gardens, Oxford — £3 post free.
2. Stephanie Kiryluk (1980), "What the Pupils Think", *Mathematics Teaching, 91*.
3. Richard R. Skemp (1976), "Relational Understanding and Instrumental Understanding", *Mathematics Teaching, 77*.

The Quadratic Equation Formula

by Russell Parker
Diss Secondary Modern Boys' School

This formula is usually developed from the generalization of "completing the square". However, there are more ways of killing a cat than choking it with cream, and no harm would be done by introducing a class to the following method, which demonstrates the important principle of *elimination.*

Suppose $ax^2 + bx + c = 0 \quad \ldots$ (i)

Now, *for all x,* $\quad x = (x - d) + d$.

Let $y = x - d$, whence $x = y + d \quad \ldots$ (ii)

Substitute (ii) in (i):

Then $a(y + d)^2 + b(y + d) + c = 0$

From which:

$ay^2 + (2ad + b)y + ad^2 + bd + c = 0$

Since d can have *any* desired value, let $2ad + b$ (the coefficient of y) equal **zero**. Then $d = -\frac{b}{2a}$ and the y term is eliminated.

We therefore have:

$$ay^2 + a\left(\frac{-b}{2a}\right)^2 + b\left(\frac{-b}{2a}\right) + c = 0$$

From which $y = \dfrac{\pm\sqrt{b^2 - 4ac}}{2a} = x - d = x + \dfrac{b}{2a}$

And hence: $x = \dfrac{-b \pm \sqrt{b^2 - 4ac}}{2a}$

Trees and Leaves

by Richard Walker

I want to describe a topic which I have found very successful with lower secondary pupils, especailly those of limited ability. The prime factors of a number are found by a slightly different method from usual. The process is pictured as the growth of a tree.

I can explain this best by an illustration. Suppose we want to factorise 24. This is the root of our tree.

Find any number, apart from 1 or 24, which divides into it: 6 for example. 24 divided by 6 is 4. The tree grows two branches.

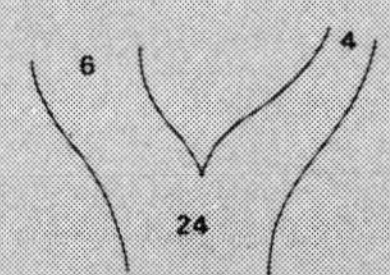

Further branches sprout from 6, because 2 goes into it 3 times. The branch ending in 4 also grows.

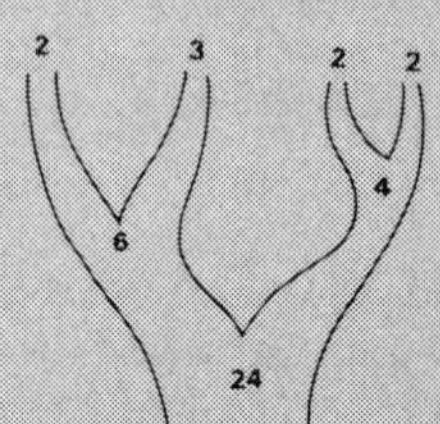

No more growth is possible, because 2 and 3 have only themselves and 1 for divisors. The numbers at the ends of branches which stop growing are the leaves.

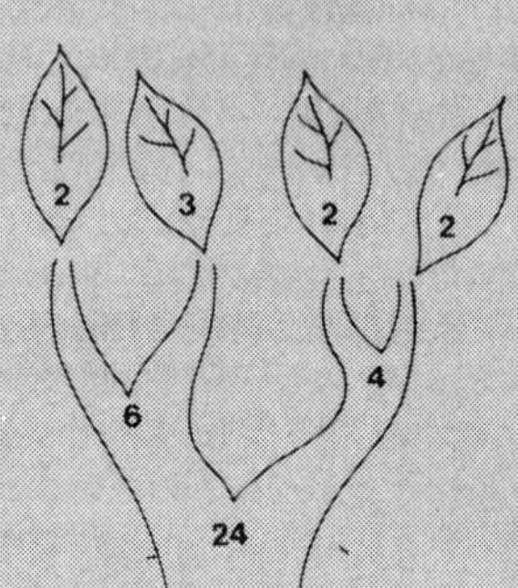

Finally we have a fully grown tree, and the leaves are the prime factors of 24.

The basic idea is that if the number at the end of a branch has proper factors, further branches sprout from it. When a branch stops growing the terminal number becomes a leaf.

Most pupils, even those not normally very successful at mathematics, are able to construct their own trees after seeing two or three examples developed. There are in general several different trees rooted in any particular number, but children seem to realise fairly quickly that all these have the same leaves.

Pupils find the drawing of these tree diagrams enjoyable and the artistically inclined have an opportunity to display their talents. Constructing the tree for a large number can present quite a challenging task. The most eloquent testimony to this and to the success of the topic is provided by the words of the small boy who exclaimed: "Here, Sir, why are you making us think?"

Why doesn't $(a + b)^2$ equal $a^2 + b^2$?

by K. Robin McLean,
Liverpool University School of Education,

It was the second session of a PGCE Mathematics for non-specialists option course. In the group of twenty there were chemistry graduates, a number of geographers fairly well-versed in statistics and some students who had done no mathematics since they took 'O'-level. I was attempting to show how algebra could be developed from number work in secondary schools when Margaret, a mature biology student, said "I know that $(10+9)^2$ isn't the same as 10^2+9^2, but why isn't it the same?"

What an opportunity for the rest of the group to try their hand! Everyone rose to the challenge, all eager to help. Soon the board was covered with explanations of all kinds. Brackets were multiplied out, other numbers were substituted in place of 10 and 9, someone based an argument on analogies with third and fourth powers, pictures of squares were drawn and areas examined. Each fresh explanation was greeted with approval by the majority of students, but Margaret's response was always the same: "Yes, I can see that. That's very nice, but I still don't understand *why* $(10+9)^2$ isn't the same as 10^2+9^2."

We broke up for coffee and, on resuming the class, Sue, a PE student, wrote on the board:

	19	10	
	19	10	
	19	10	
	19	10	
	19	10	10^2
	19	10	
	19	10	
	19	10	
	19	10	
19^2	19	10	
	19	9	
	19	9	
	19	9	
	19	9	
	19	9	9^2
	19	9	
	19	9	
	19	9	
	19	9	

"Obviously", said Sue, "If you add up the left-hand column you'll get more than if you add up the right-hand column for, in each row, the number on the left is bigger."

Silence fell. That said it all. Margaret was satisfied.

IRRATIONAL LEGO

by Alan Barnes and Barry Martin,
The University of Aston in Birmingham

Fig. 1

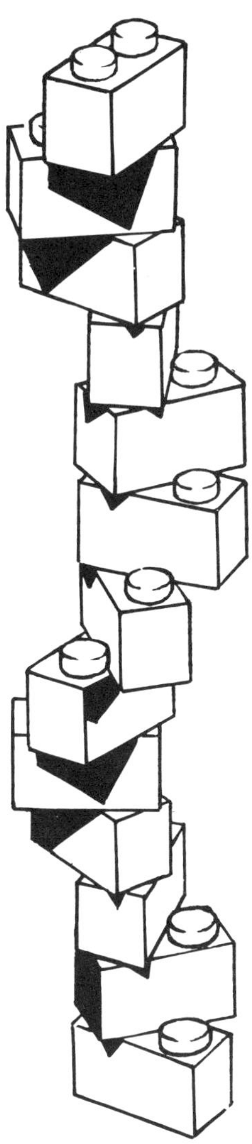

Mr K. D. Jeffs of Waterlooville noticed that if the simplest Lego bricks are stacked as shown (Fig. 1) and the column gently twisted so as to take up the slack the structure appears not to repeat.

The dimensions of the bricks, kindly supplied by Paul Møller of Lego Systems, Denmark, are given in Fig. 2 where two are shown in the relative position that they occupy in the column.

If the structure is to be periodic we must have integers m and n such that

$$n\theta = 2m\pi \text{ and } \mathrm{Sin}\theta = \frac{637}{800}$$

However, it is known[1] that if θ is a rational multiple of π then Cosθ is irrational except when Cos$\theta = 0, \pm \frac{1}{2}, \pm 1$. From the identity

$$\mathrm{Cos}\left(\theta - \frac{\pi}{2}\right) = \mathrm{Sin}\theta$$

it follows that the only rational values of Sinθ when θ is a rational multiple of π are again $0, \pm \frac{1}{2}, \pm 1$, hence the Lego column is non-periodic and provides a nice realisation of an irrational sequence.

Incidentally $3\theta = 158.3..^\circ$ which makes it possible to build a passable regular 17-gon($158.8..^\circ$) with Lego bricks; they are just deformable enough to accommodate the half degree discrepancy.

Fig. 2

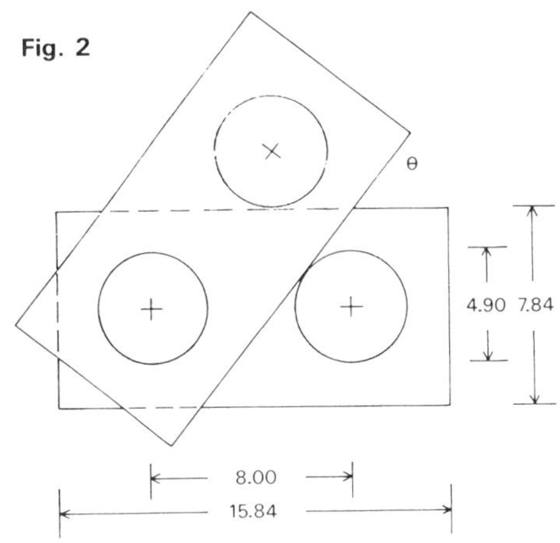

Reference

1. Srinivas, V. Regular polygons on a square pinboard, *Mathl Gaz.*, **61**, 290-2 (No. 418, December 1977).

4 Calculators and Computers
by Ken Brown

The introduction of new technology into the mathematics classrooms of Britain has been complemented by the development of new teaching strategies designed to utilise it effectively. Developments have been reflected by the many excellent articles within the pages of *Mathematics in School*. The small selection here is intended to give a flavour of the enormous variety of activities that are possible using computers and calculators in school mathematics.

With the rapid pace of change some articles in this area date quickly. Such dating is inevitable when one considers the rapid drop in the price of machines coupled with the increase in calculating power available. I can remember being a pupil at school not too long before the first issue of *Mathematics in School* was published and being shown the mathematics department's calculating machine. It was an exotic mechanical contraption consisting of several levers used to set numbers and a handle which had to be turned to complete the calculations. We were, I recall, allowed to attempt one or two 'sums'. Finding myself possessing neither the manual dexterity nor the stamina to cope with the endless cranking of the handle, I marvelled at the ingenuity of humankind to invent a device that actually made arithmetic not only mentally tiring but physically exhausting too!

But how the outlook has changed! Most primary schools now have sets of calculators and we have national projects such as PRIME which investigate the role of calculators in mathematics education. Computers are commonplace in schools and they are becoming more powerful and easier to use.

The present low prices of calculators are highlighted by the *Apparatus Review* of Volume 1 No 6 where a very basic electronic calculator cost about £50. Today the calculator is an everyday part of people's lives, nearly everyone has access to one. The price of computers continues to fall.

Both the calculator and computer are now an established part of mathematical education and I am confident that we can look forward to their use becoming an integral part of many more mathematics lessons and articles in *Mathematics in School*.

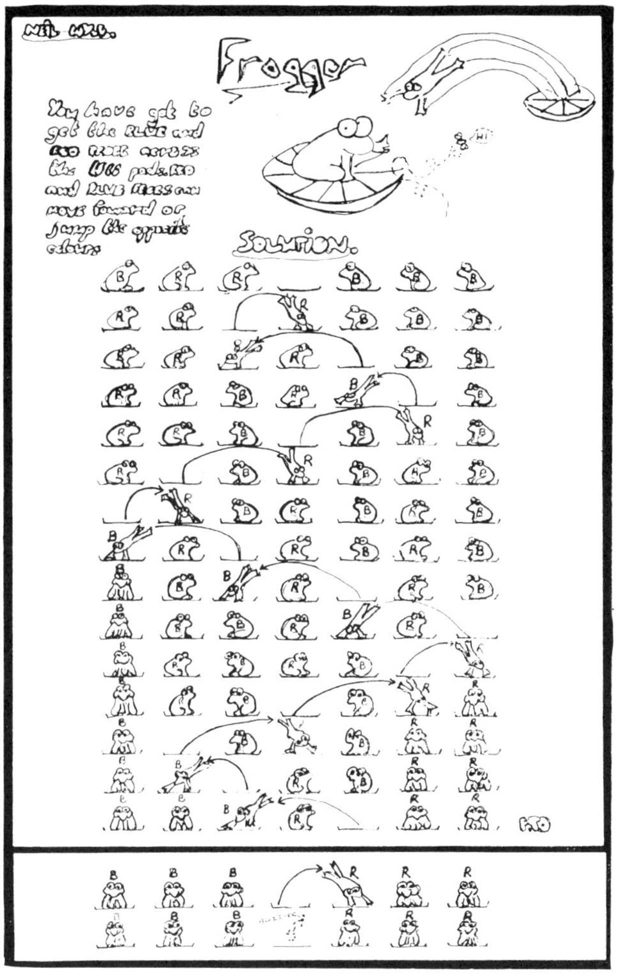

Procedures and Algorithms in Primary Mathematics

by

Anita Straker

"Well', said Owl, "the customary procedure in such cases is a follows."
"What does Crustimoney Proseedcake mean?" said Pooh. "For I am a Bear of Very Little Brain, and long words bother me."
"It means", said Owl, "the Thing to Do."

(A. A. Milne in Winnie–the–Pooh, 1926)

Al–Khowarazmi was an Arab mathematician who was associated with the development of our decimal system of numeration. His name is the origin of the word algorithm, originally used to describe an arithmetical process, but nowadays used more broadly to describe any logical procedure which will solve a defined class of problems.

All primary teachers will be familiar with at least four algorithms: the rules for adding, subtracting, multiplying and dividing two decimal numbers. These algorithms are very powerful. The decomposition rule, for example, works in any circumstance when one decimal number needs to be subtracted from another. But although the four rules have wide application, there are times when they are cumbersome to use. Just consider what happens when the decomposition method is used to calculate the difference between 10 000 and 4 837!

A great deal of time has been spent in the primary school helping young children to understand and learn difficult rules like the decomposition method, as though such algorithms had intrinsic value in their own right, rather than being — as they are at this stage — a means to an end. Procedures and algorithms are certainly important in mathematics, but do primary children have to start with a study of those which are particularly difficult?

The Cockcroft Report suggests that problem solving is at the heart of mathematics, and at the heart of problem solving lie algorithms, for in solving problems an algorithm is developed — a procedure which can be followed in order to obtain solutions to problems of a particular kind. This is a powerful argument for learning more about algorithms in mathematics. But once a precise algorithm has been defined, it is often possible to program a calculator or computer to carry it out. The fact that these machines are now common place in schools makes the argument even stronger.

> "In the light of advances in technology, primary and secondary school mathematics should include the explicit study of algorithms as an integral part of the curriculum. This means that all pupils will need to be involved in all stages of the study i.e. the performing of given procedures, the modification of procedures, the design and evaluation of new or alternative procedures. This requires the availability of suitable tools from pencil and paper and concrete apparatus to calculators and computers, and a full appreciation requires that pupils should make some use of a suitable programming language... e.g. BASIC or LOGO."[1]

If a study of algorithms is to become an integral part of the primary school curriculum, what should it involve? It may come as a surprise to primary teachers to find that in many aspects of their day to day work in the classroom children are learning and using procedures and algorithms long before the four rules are encountered. From birth — and some would argue even before this — we have been devising algorithms in order to make sense of and control our own world. The fact that nearly all children can talk is one example which illustrates that they can identify and use algorithms naturally. For although most babies utter sounds readily enough it is only when the sounds are combined according to certain accepted rules or conventions that they become meaningful and useful.

The first aim in school might therefore be to help children become more aware of the algorithmic processes which they are already using. The eventual aim might be to ensure that they can devise procedures or algorithms for themselves, and define them in a precise manner. The stages which children pass through might be:

1. **able to follow or to use a procedure defined or described by someone else;**
2. **able to describe familiar procedures in an informal way;**
3. **confident at making modifications to procedures to do different tasks;**
4. **able to make up their own procedures;**
5. **able to describe or define a procedure in a sufficiently precise manner so that someone else can carry it out;**
6. **able to explain how someone else's procedure works;**
7. **able to compare and evaluate different procedures;**
8. **able to modify a specific procedure so that it can be applied to the general case.**

The primary school curriculum is already overcrowded. If it were necessary to introduce totally new activities in order to help children to develop these skills, then some heart searching would need to take place. Fortunately, most can be realised by giving a different emphasis, or order of priority, to some of the activities and experiences which are already a normal part of a school day.

Coming to School

Days and Nights

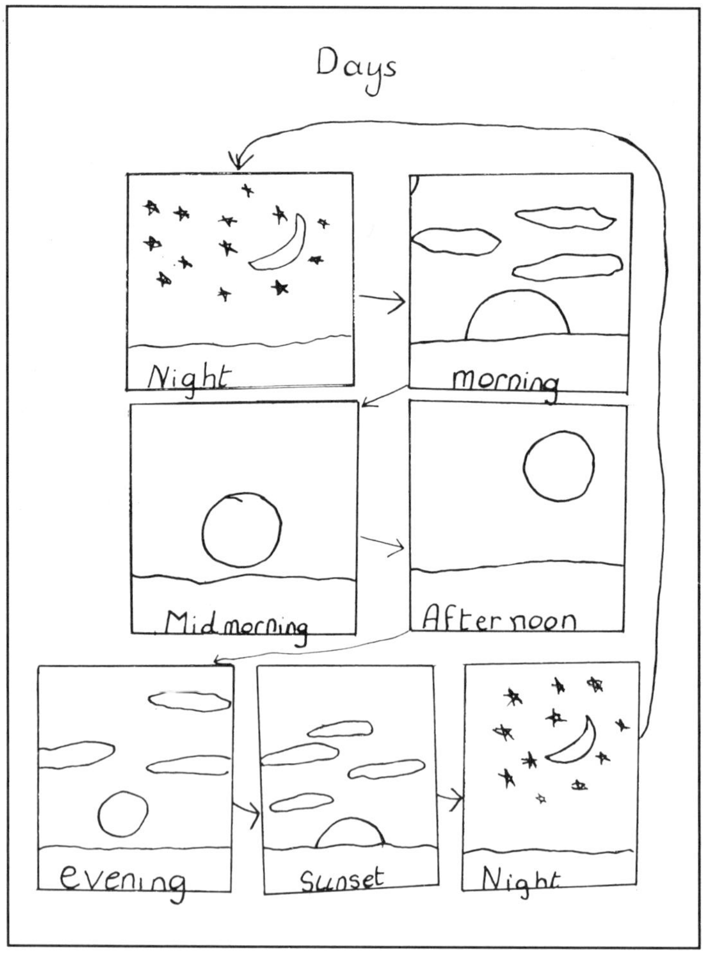

For example, most infants should be able to identify the stages in ordinary processes like coming to school, getting out of bed and getting dressed, laying the table for a meal, using the library to borrow a book to read. The stages can be illustrated by an ordered sequence of mimed actions, a set of pictures, or a list of simple sentences, and the illustrations can be given to other children to interpret. Some of these procedures are cyclical — and so children can invent their own way of indicating that once the process is finished it should be repeated.

In their PE lessons infants will learn body movements or simple dances and play games like Simon Says, making use of instructions like forward, backward, left, right, turn, roll, hop, often with numbers attached to them: go forward three steps. Again, children can make up their own sets of ordered movements, and find ways of signalling these to each other, perhaps using percussion instruments.

Infants are also accustomed to following sets of instructions to make patterns with beads, with felt shapes, or with pegs on pegboard. Making up coded sets of instructions for other children to carry out could be their first experience of writing a program: 3R, 2B, 1R, 2G for three red beads, followed by two blue, one red, two green. And what kind of code should be used to show that this pattern should be repeated six times?

As they carry out these sequencing activities, infants will start to develop their own form of an algorithmic language — something a bit like a programming language, but much more loosely defined. The vocabulary is likely to include words like: first, last, after that (or next), jump on, jump back, once more, start, stop, do again (or repeat), until, if, then, otherwise (or else), finish (or end).

This might be the appropriate time to introduce activities with programmable toys like George or BigTrak. BigTrak is a battery–driven robot (now, alas, hard to obtain), disguised as a white moon buggy. With a little ingenuity programmable toys like BigTrak can be co-opted into the children's interest topic, by being decorated to look like anything from a green frog with ping–pong eyes, to a sailing ship from the South China Sea.

Such toys can be operated so that they carry out one action at a time, but they can also store a complete sequence of instructions to be carried out continuously. Activies with BigTrak can of course be suggested by the teacher, but the best suggestions come from the children themselves. "How can we teach BigTrak to ...

go to the top of the slope and stop;
move in a circle
go all round the table and back to us again;
zig–zag through the bean bags;
make the waves on the sea;
do a dance with the other BigTrak."

A fly–on–the–wall listening to a group of six year olds working unsupervised with BigTrak is likely to hear some interesting and spontaneous snippets of conversation:

Go left a bit more.
Try (a number) and see.
First go forward, then right.
Next, we'll have to...
Do it again.
If Big Trak does that, then it will have to ...
Let's see what happens if ...
How many more times?
Try a bigger/smaller number.
Let's try changing...
Yes, we've done it!
Come and see what we've done.
How did you do that?
Can I try that?
Will you show me?

In the junior school, the infant activities can be extended, and made more challenging. many different activities offer an opportunity for following instructions, for giving instructions for others to carry out, and for finding informal ways of recording these.

Programming Big Trak

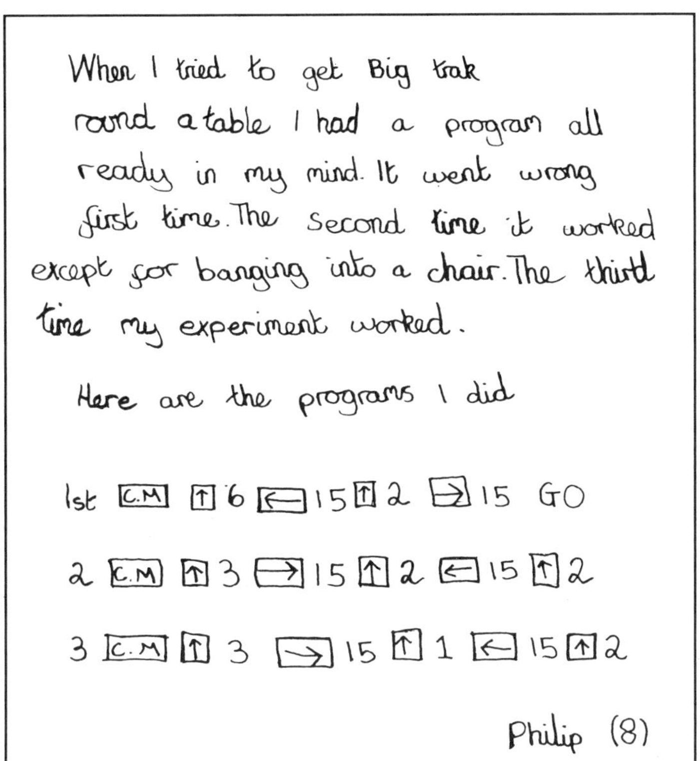
When I tried to get Big trak round a table I had a program all ready in my mind. It went wrong first time. The second time it worked except for banging into a chair. The thirtd time my experiment worked.

Here are the programs I did

1st C.M ↑ 6 ← 15 ↑ 2 → 15 GO

2 C.M ↑ 3 → 15 ↑ 2 ← 15 ↑ 2

3 C.M ↑ 3 → 15 ↑ 1 ← 15 ↑ 2

Philip (8)

A Big Trak Program

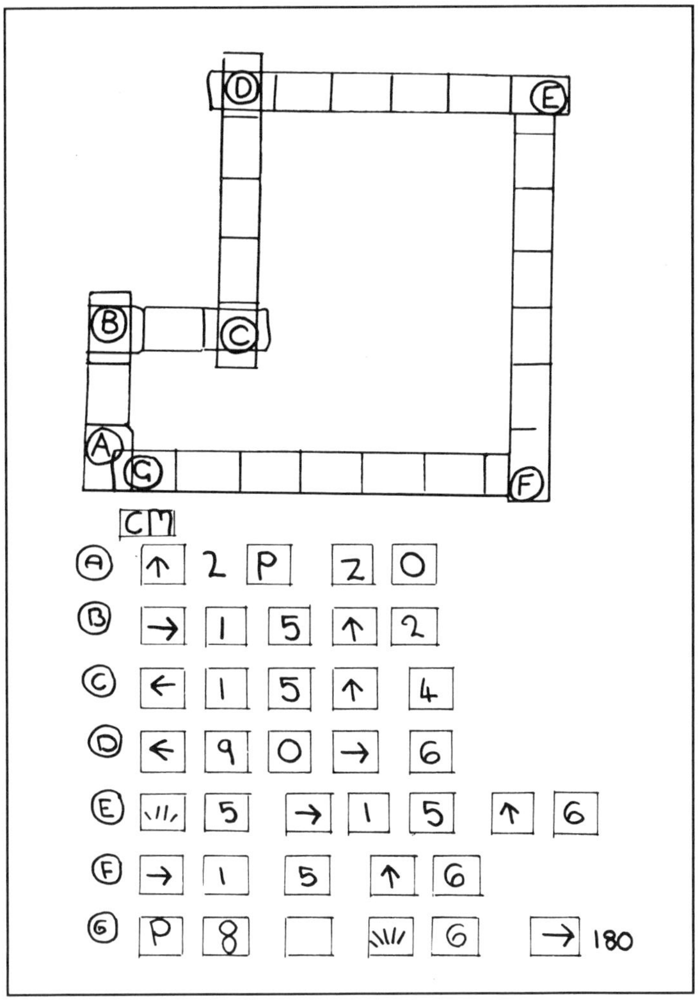

Children need positive encouragement to look for, modify, and them make up their own real–life 'programs': rules for games, recipes, instructions for model making, for controlling traffic lights, for ringing peals of bells. Some of these procedures may have branching statements in them, or loops or repeats; some, like knitting patterns, country dances, or musical scores, are coded.

This kind of work can be complemented by the use of puzzles in which certain combinations of moves will produce a satisfactory solution to the problem: playing pegboard games like Frogs or exploring kinght's moves on a chess board. Sometimes computer simulations of logical puzzles involving 'moves' can be found, e.g. finding a procedure which will:

make the lorry carry away the elephant in JUMBO;
steer the boat safely through the lock in LOCKS;
put the collars on the dragons in GRANNY'S GARDEN;
lift the bales off the dock-side and on to the boat in CARGO;
find the shortest route through a maze in CRASH;
pour out a particular quantity using JUGS;
ferry people and/or animals across a river in BOAT or in FARMER.

In these programs there may be several possible sequences of commands which will produce a satisfactory result. But often there is little emphasis on recording or coding the sequence of moves, of comparing one satisfactory procedure with another, or of refining a successful procedure, unless the teacher encourages the children to do so. If children are to become successful programmers these are important skills.

How to Swim

1 Go up stairs in your house.
2 Get your swimming costume.
3 Get a towel.
4 Go down stairs.
5 Get into car turn key and drive to pool.
6 Park car in car park
7 Walk into Sportscentre.
8 Go up stairs.
9 Go to changing room and get on costume
10 Put clothes in locker.
11 Take out key and put round wrist.
12 GO through shower.
13 Go to deep end.
14 Say super man dive in and swim.

Thats how you Swim

Michael (9)

An Evening at Home

Solution to the Farmer's Problem

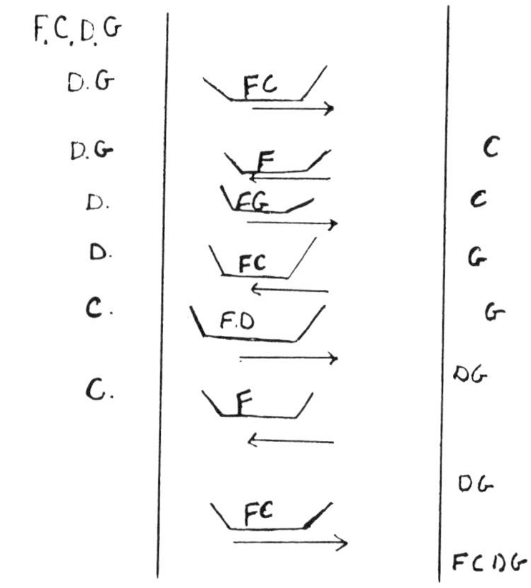

The program farmer was very good because I liked the way the farmer went across the river. I did like the dog because of the shape of it. To improve farmer I would have some boats so that you can dogde them.

Boat

This is a report of a computer program called "BOAT". You have to try and get two women and two girls across the river in a boat and no-one can swim. You are only allowed 1 adult in the boat, one child in the boat and two children in the boat. On the easy level there are two women and two children in the boat. The most least times the boat can cross the river is 9 and this is how you do it.

THAT'S IT

To do the next harder puzzle do this sequence twice. By Graham Easton.

There are number situations too which require an algorithmic approach. This example is taken from *Limerick Number Puzzles* by Gregory and Seymour.

> I'm a two digit prime and no more.
> My digits subtracted is four;
> Their sum is not ten.
> I say to you then,
> To find me is not a big chore.

One algorithm for a solution might be:

> Ordered list: two digit numbers
> Delete: multiples of 2, 3, 5, 7
> Table: differences of the digits
> Delete: any prime for which the difference of the digits is not equal to 4
> Table: sums of the digits
> Delete: any prime for which the sum of the digits is 10
> Answer: 59

Number work of this kind needs to thread throughout the junior years. There are plenty of ideas to choose from. How many different procedures is it possible to invent:

- for subtracting 37 from 92;
- for multiplying 33 by 5;
- for making 83 out of 4s and 5s;
- for adding up all the whole numbers from 1 to 20;
- for making different arrangements of the digits 1234;
- for guessing a number in the 1 to 100, when the clues are 'too big' or 'too small';
- for testing whether a number is divisible by 12;
- for pouring out 1 litre of cider when only a 7 litre and a 5 litre jug are available;
- for shading three–eighths of 4 × 2 rectangle?

Structure Diagram for Knitting a Jumper

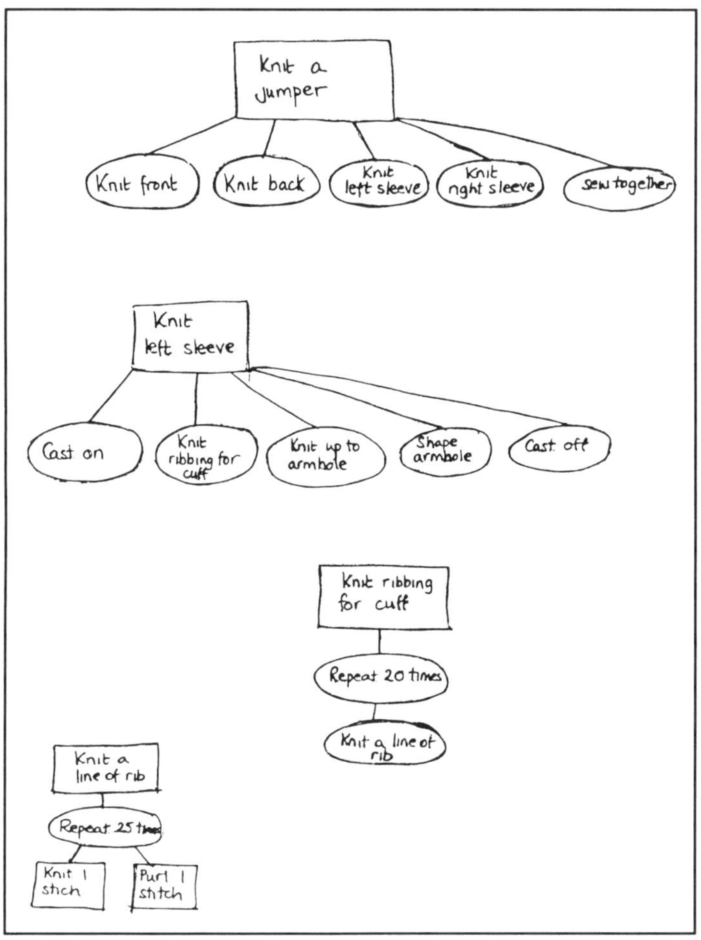

The procedures or algorithms which produce the solution to both the logcial and the numerical problems might be represented in various ways. A more permanent record often helps children to evaluate their algorithms, to decide whether they are efficient, easily understood, and even elegant. The representation might be a set of pictorial instructions, a list of written or of coded instructions, a flow chart, or — perhaps more appropriately in a primary school — a structure diagram. (Structure diagrams were invented by R. W. Doran who wanted to find a way of helping children to see how an algorithm can be analysed into smaller and smaller independent component parts until it is broken into basic elementary instructions.)

Target Game

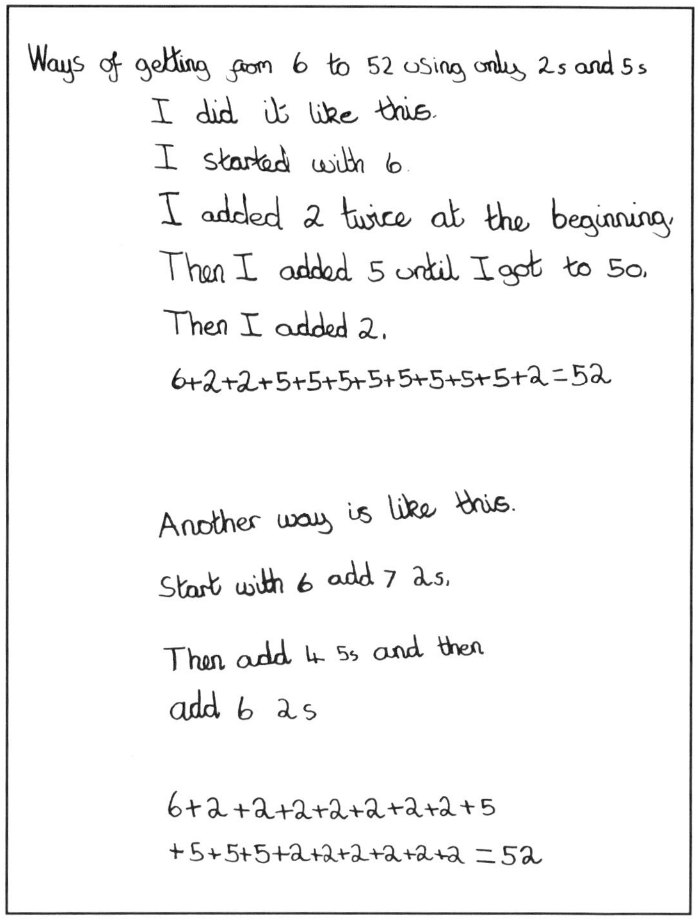

Many of the activities with numbers which lend themselves to an algorithmic approach will involve the use of a calculator. Infants might be already using a simple yet fundamental algorithm to add two small numbers together: enter a number, press +, enter another number, press =. (This algorithm, like the old four rules algorithms, is a powerful one. It applies not just to the addititon of small numbers, but to all numbers within the range of the calculator.) Later, children will use combinations of similar processes in target games, where one number is entered in the display and ways of arriving at another number have to be found. For example, how many different ways are there of starting with 4 and reaching 23 using only 2 or 3 and the + key?

At the junior stage there needs to be considerably more emphasis on the algorithms which can be developed and used with a calculator. There are those which make use of the constant for repeated addition or multiplication, rules to generate sequences of triangular numbers, square numbers, Fibonacci. Some children may enjoy being given an algorithm to experiment with, perhaps for finding a square root, a cube root, or the highest common factor of two numbers. Satisfaction that the algorithm is effective leads inevitably to the queation "Why does it work?"

Flow chart for finding a cube root on a calculator

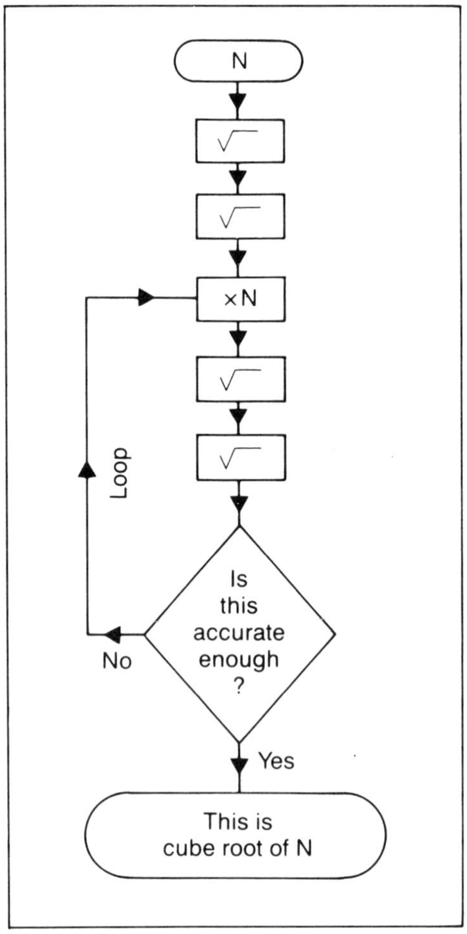

Why does it work?

Computer programming — working out an algorithm to solve a problem and then teaching a computer to carry it out — is a mathematical experience for children of all ages. It starts with the introduction of the simple commands needed to make a turtle move, FORWARD, BACKWARD, LEFT, RIGHT. Nearly all children seem to enjoy producing an effect using a sequence of these commands, and will discuss and plan their sets of instructions even before they start to work with the micro.

Named Procedures for DART

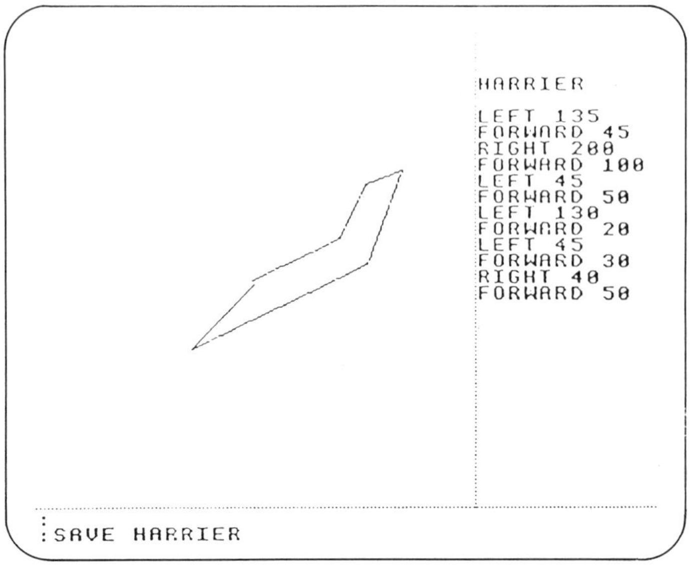

After some time spent controlling the turtle by driving it directly, they can be encouraged to collect, name and save a satisfactory group of such instructions as a procedure which can be used on another occasion. They can then spend some time making changes to their own and other children's procedures to see what happens, and can try combining procedure in various ways.

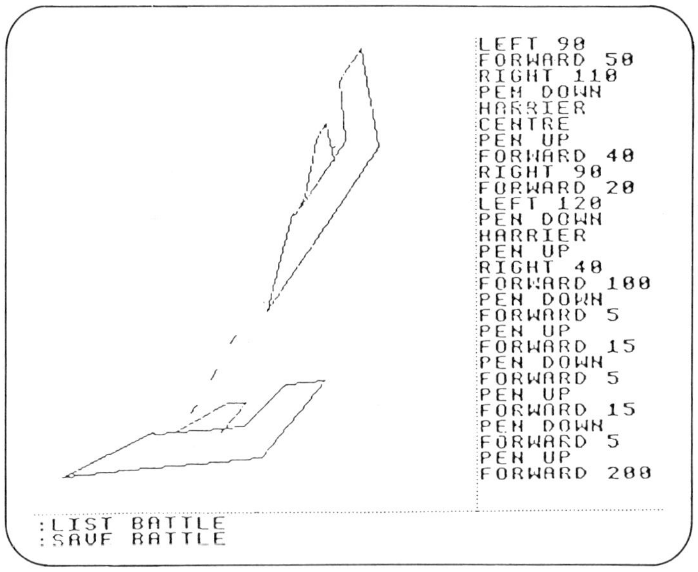

Not all the programming which primary children undertake involves graphics. In some primary schools children are already using simple commands like SWITCHON or SWITCHOFF to control a working model which they themselves have designed and made. In other schools children are making up programs which will calculate (multiply one number by another, find the average of three numbers, find the perimeter of a rectangle). Such procedures can readily be altered (to divide one number by another, to find the average of four numbers, to find the area of a rectangle), and some primary children will be able to generalise them (find the area of any rectangle, the average of n numbers, the perimeter of a polygon with n sides).

Cube Program

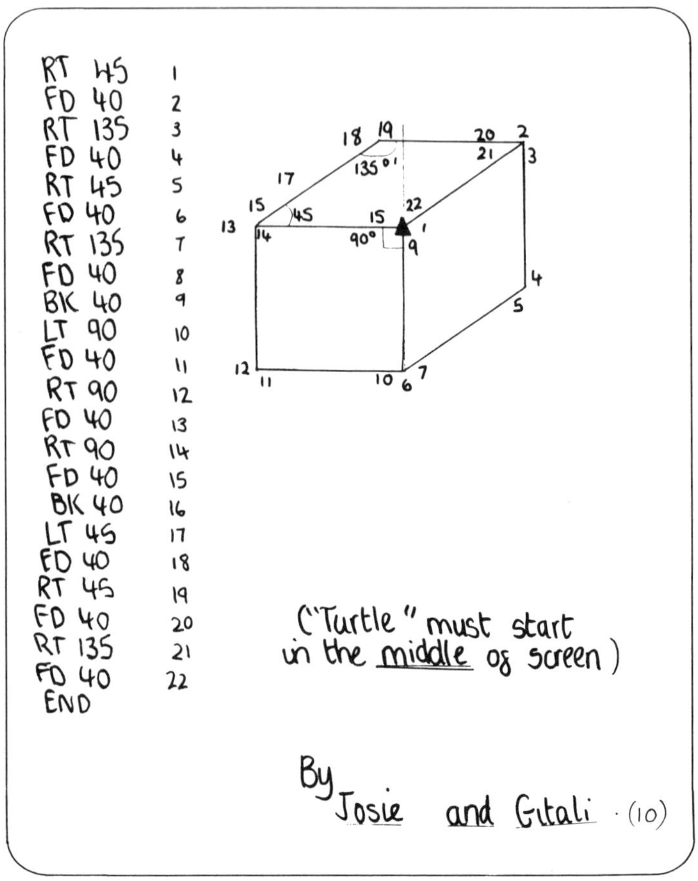

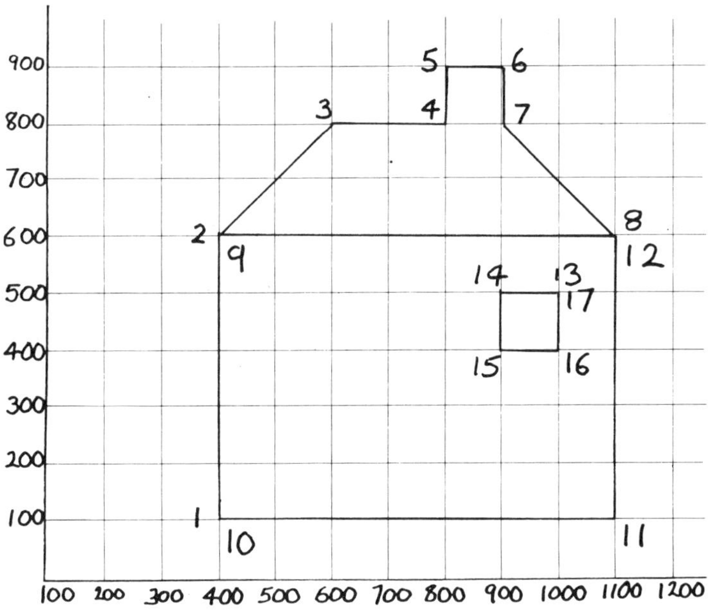

House Program in BASIC

created by nine year old children working together in a small group

```
1 MOVE 400,100          10 MOVE 400,100
2 DRAW 400,600          11 DRAW 1100,100
3 DRAW 600,800          12 DRAW 1100,600
4 DRAW 800,800          13 MOVE 1000,500
5 DRAW 800,900          14 DRAW 900,500
6 DRAW 900,900          15 DRAW 900,400
7 DRAW 900,800          16 DRAW 1000,400
8 DRAW 1100,600         17 DRAW 1000.500
9 DRAW 400,600
```

Rectangle Program

My Program for a rectangle

10 PRINT "what is the width "

20 INPUT width

30 PRINT "what is the Height "

40 INPUT ~~HEIGHT~~ height

50 Perimeter = 2 * (width + Height)

60 PRINT "Perimeter is"

70 PRINT perimeter

80 Area = width * ~~HEIGHT~~ height

90 PRINT "Area is"

100 PRINT Area

Susan (11)

In some of these generalisations a process may need to be repeated, and at the same time other investigstions which involve repetiton or recursion can be encouraged: inventing and modifying an algorithm to order the numbers in a list, to draw spirals, to produce number chains, to control a model of a level crossing.

> Problem: To order a list of 6 numbers e.g. 3 2 6 5 1 4
> Permitted move: Reverse any number of digits in the list, counting from the left.
> Find the position P of 6, the biggest number.
> Reverse P numbers (to put 6 at the front).
> Reverse 6 numbers (to put 6 at the back).
> Now repeat for a list of 5 numbers, and so on.

Before calculators and computers were commonp'ace it may have been reasonable for primary children to concentrate on understanding and applying the four rules, powerful but difficult algorithms developed by others. Nowadays, except in the simplest of cases, the peformance of arithmetical algorithms is done by a machine. We can no longer justify skill training in the efficient performance of pencil and paper methods for long multiplication and division, for operations with fractions, for the calculations of percentages. Instead, we need to encourage children to invent their own procedures for solving problems, and help them to make productive use of the technology which will be commonplace in 10 years time.

There are at least six (and probably more) different algorithms which will produce the total cost of ten pounds of fruit for the freezer at 75p per pound, with VAT at 15% to be added. They will all work effectively. If children leave school able to 'invent' any one of them for themselves, and able to use a calculator or computer to perform it, then a great deal will heve been achieved.

> "So", said Owl, "it's important to learn about algorithms."
> "But I could never remember all these Algy Rhythms", said Pooh.
> "Can't you write then down for me?"
> Owl sighed. "You don't have to remember them."
> Pooh looked puzzled.
> Owl, trying to look confident, announced,
> "You can always make up your own."
>
> **(B. B. Milne, 1985)**

References

Mathematics Counts, Report of the Cockcroft Committee. HMSO. 1982.

Mathematics and Microcomputers: the Pendley Manor Report. *Mathematics in School.* March 1985.

Limerick Number Puzzles. Gregory and Seymour. *Tarquin (for Creative Publications Inc).* 1978.

Structural Programming. R. W. Doran. *Computer Education.* October 1971.

Note

This article was written by Anita Straker, with assistance from Katrina Blythe, Jim Edwards, Hilary Shuard, Mike Wardle, John Warwick, and Joe Watson, It stemmed from discussions in one of the working groups at a Mathematical Association conference on Primary Maths and Micros. The conference was funded by the DES, and took place at Leicester in December, 1985.

CALCULATOR GAMES

by Gordon Haigh, Elmore Green Primary School
Andy Bailey, West Midlands Co

Here are three calculator games that have been used successfully in infant, junior and secondary schools. The place of games is discussed elsewhere in this magazine. The place of calculators in the games shown here is illustrated by this diagram:

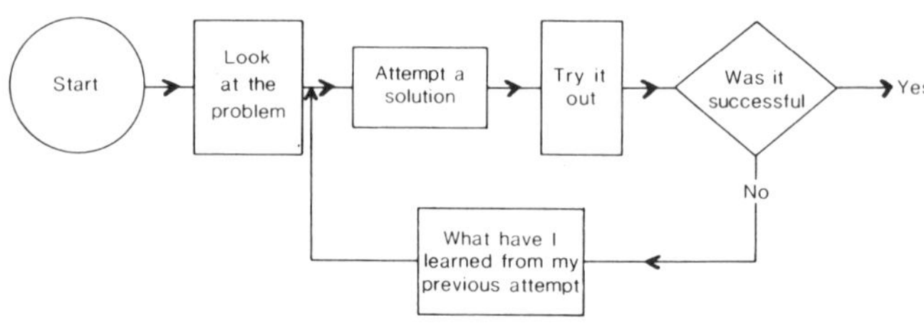

We hope to show that within this the calculator plays an important facilitating role. It promotes a "guess then adjust your guess in the light of your mistake" strategy. The mistakes become clues leading to solutions rather than the measure of suitability to belong to the human race that they so often seem.

All of the games can be played solo with minor alterations to the objectives. They then become "competitions against yourself" without the sometimes aversive problems competition can bring. In this form they are open to group solutions in which teacher can participate.

In the Infant Classroom

In a corner of a top-infant classroom four children were working in pairs around two large boards, one of which is illustrated below:

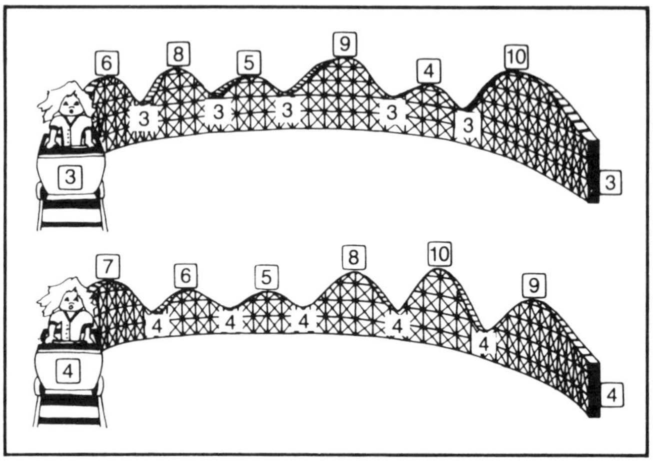

The obvious similarities between the activities of these children and that of four children in a junior classroom nearby playing a game called "Missing Links" were the use of calculators in each activity and the lively, purposeful activity which was being generated. The children were obviously having fun. Interest and enjoyment are essential requirements for creating and developing attitudes towards mathematics but in assessing the value of an activity there must be other elements present for it to be fully worthwhile.

The following list of criteria for the mathematics curriculum reflect recent circulars and reports:

1. *learning should be interesting and enjoyable, developing healthy and positive attitudes,*
2. *there should be a correct balance between oral, mental and mechanical work,*
3. *opportunities should exist for investigation and for recording these in a relevant manner,*
4. *children should be encouraged to think logically,*
5. *the skills of approximation, estimation and estimate and check should be developed,*
6. *opportunities should be provided for searching for and appreciating pattern,*

Using this list it may be helpful to examine the games being played by these children to see to what extent they fulfill these criteria.

The game played by the young children involved moving a token along a track as a reward for performing simple operations using a calculator, the first player to get their token to the end of the track being declared the winner. Daljit for instance starts with her token on 3 and according to the rules of the game has to enter this number into the calculator and change it to 6 (the next numbered stage on her track) without using either the off or cancel keys. If she does this correctly she will be able to move her token to the next stage. She therefore has to solve the problem

3 for — □ = 6

First Daljit must decide whether 3> or <6. Then in deciding 3 < 6 the next question is — by how many? This resolved she can then key in the following — 3 + 3 = 6. Having achieved her target number she moves her token along the track.

Her opponent Jenny starts at 4 and has to aim for 7. Jenny's solution was to key in 4 + 2 = 6. According to the rules of the game she had to leave her token where it was and attempt a correct solution to the problem on her next turn.

On her next turn Daljit enters 6 into the calculator and asks herself whether 6> or <3 which is the next numbered stage along the track. As the answer is > she decides that this time she requies the − operation and performs the correct operation 6 − 3 = 3 and delightedly moves her token along the board.

Jenny tackles her next turn with greater conviction and keys in the correct solution 4 + 3 = 7, claps her hands and moves the token.

On the face of it the children are engaged in an activity which is practising number bonds and using addition and subtraction facts. To suppose that this was all there was to this activity would be a rather superficial view which reference to the oral exchange between the girls at this point of the game may go some way to revealing.

Daljit *"I've got to go up"*.
Jenny *"Up to six"*.
Daljit *"It's add. I've got to add ... something to three. I know, I know.* (speedily keys in correct solution and looks very pleased with herself)
Jenny *"Mine's harder. Look I've got to get higher to seven"*.
Daljit *"Press add — it's there.*
Jenny *"Add two ... Oh!"*
Daljit *It's not enough. You didn't add enough"*.
Jenny *"I know"*.

There is interaction between the two girls. Daljit is supporting Jenny by providing her with hints and clues. At a very simple level Jenny's effort aided by Daljit uses a problem solving approach looking at her initial error to provide an eventual solution.

It was Kevin one of the two boys playing the same game who after some 10 minutes said, "*the take-aways are easy*". "*They're not*", retorted Jenny who was obviously experiencing some difficulty with this operation.
"*They are*", persisted Kevin and invited by the teacher he began to try to explain to the others the relationship which he had discovered. (Taking 3 as the starting number) "*If you add three to make eight so you must take off three to get back*" (to 3).
"*Is Kevin right?*" the teacher asked the others. Daljit thought he was, the others were less sure so at the teacher's suggestion they set out to check Kevin's idea by recording the number sentences which were moving them along the track i.e. the successful ones. As a result of doing this it became clearer to the children that this was the case. Questions such as:

"*does it work with any numbers?*"
"*does it work with bigger numbers?*"
"*does it work the other way round?*" (if $a + b \Rightarrow c - b \Rightarrow a$ will $c - b \Rightarrow a + b \Rightarrow c$?)

were all referred back to the children by the teacher for investigation with the help of their calculators. Further play was suspended quite voluntarily while the children investigated these suggestions and confirmed the inverse relationship between addition and subtraction.

Many of the calculations being performed could have been done without the use of a calculator. It's value was in providing a check and in the case of a competative game an arbiter (a very useful function with younger children). It gave the children confidence to try out specific examples encouraging estimate and check strategies and was particularly useful when the children went on to confirm Kevin's discovery using larger numbers. It allowed their attention to remain on the real object of the activity which was discovering and confirming a relationship by relieving the children of the burden of calculation while at the same time generating a large number of examples.

Games are often used to consolidate skills and routine and this could have been an observer's initial impression of this particular activity. In this instance however, the value of a game as a lead into a particular teaching point had been noted by the teacher who had used this game as part of a deliberate teaching strategy which can be illustrated thus:

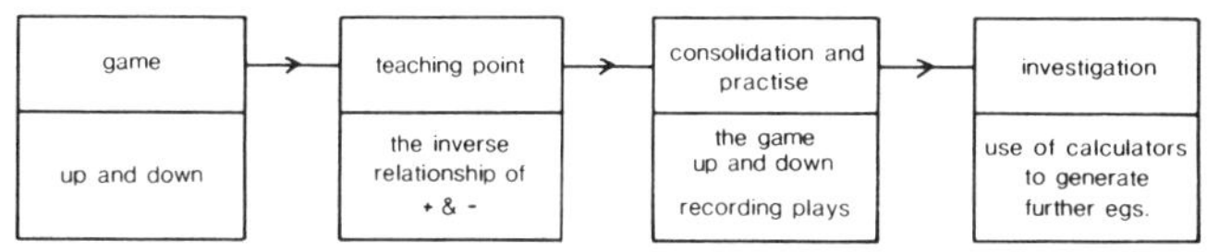

In the Junior Classroom

The four junior children aged 10 and 11 were playing a game called Missing Links, the aim of which is to construct a sequence of numbers which have a relationship to each other and to present this to others in a partly complete form to be filled in. Nasir and Andrew had made a sequence starting with 32 where each subsequent number was the sum of $\frac{1}{2}$ its precedent plus the precedent.

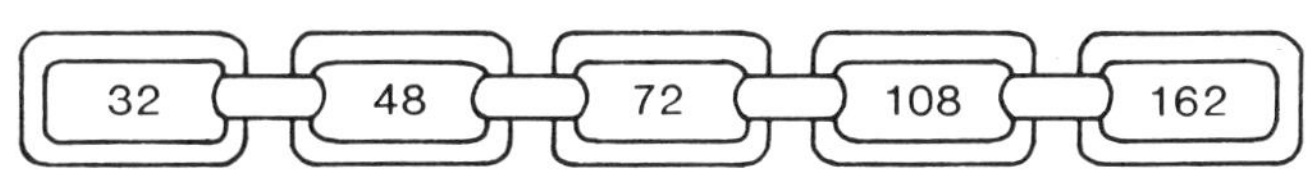

This was presented to Katie and Parveen as:

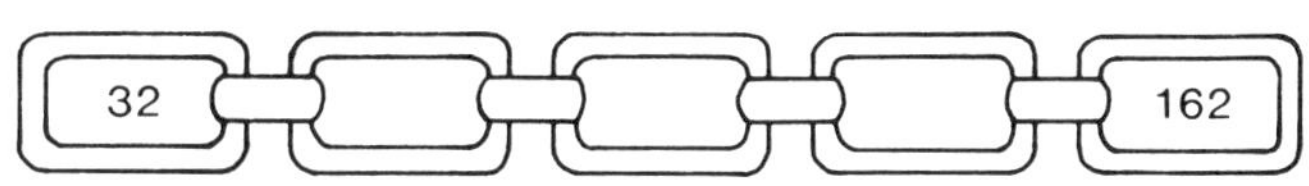

The girls started this round of the game with 20 points from which points are deducted each time a permitted form of question requiring a yes/no answer is asked, or an incorrect number is suggested, or else for time faults. On receiving an afirmative reply to their enquiry as to whether the first missing link is between 40 and 50 the girls narrowed the field to between 45–50. They then consulted for some moments before suggesting 48. When asked why they chose this they answered that it was a member of the eight times table as was the first number in the sequence. It was a logical suggestion but although it saved needing to ask more questions it would prove a red-herring if it continued to be applied as Katie realised when she used her calculator to check whether 162 was a multiple of 8. Parveen, however, was not to be deterred from this line of thought suggesting 64 as the next link and then 72 which proved correct. It was at this point that Katie exerted her influence to pursuade Parveen of the futility of pursuing a line of enquiry, that had so far brought them success, any further. She used the calculator to back up her argument. After a further consultation using the calculator to check suggestions the girls eliminated any relationship with the 4 × table and went on to investigate other possibilities.

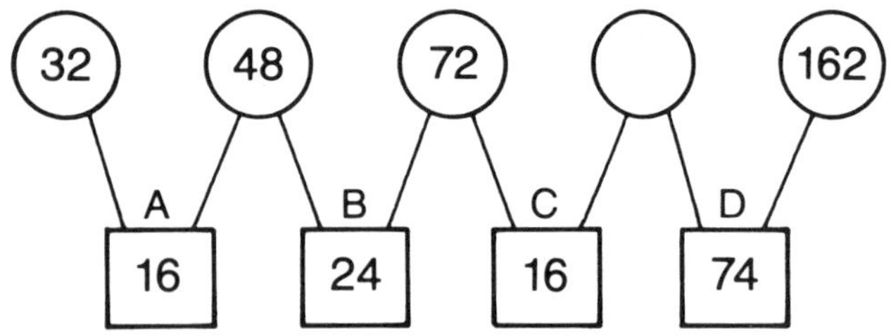

"This will not work. The difference in the box D should be 24."

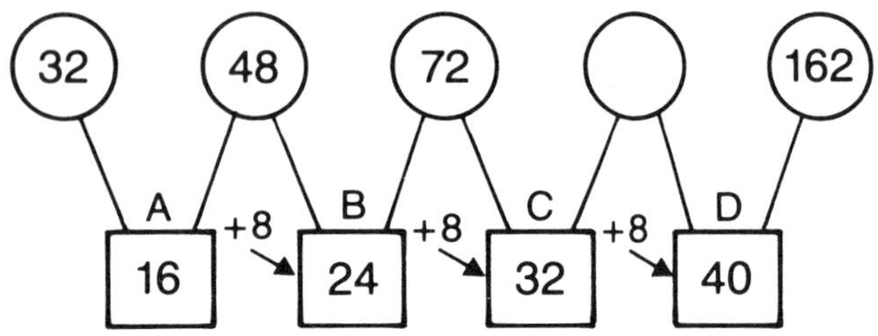

"If we add 40 to 104 the answer for the last link will be 144 we know it should be 162."

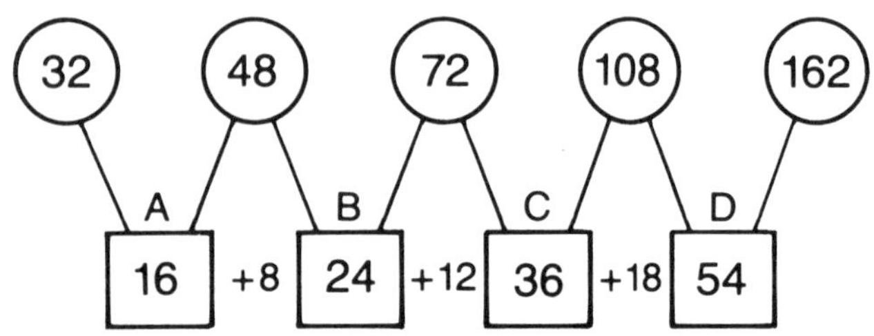

"The pattern is made by making the differences between the difference boxes greater each time by two if the chain was longer the next difference would be 80 so the next number in the chain would be 242."

Readers will be well aware that these children were quite familiar with this game and had done a considerable amount of work on patterns.

Although the children had not identified the pattern used by Andrew and Nasir to construct their chain they had nevertheless systematically devised a pattern of their own which produced a positive result.

In considering once again the criteria for the mathematical curriculum which were listed earlier we would hope that these case studies would demonstrate the scope calculator games can provide for achieving all six. Should further proof be required we suggest you try them out with groups of children of suitable ages or ability.

In the Secondary Classroom

The game described here and its many variations all pose problems of the type

$$73 \times \boxed{?} = 46$$

They do not offer "conceptual handles" to aid understanding but, by much feedback, offer a rule discovery environment in which strategies can be developed.

It is easily adjusted to suit different abilities and I have had most success pairing children of similar attainment and asking them to play it co-operatively initially.

Ncroach

A. This game has many variations. They will be explained after an example. The example has worked well with middle ability 14 year olds.

B. 2 players Red and Blue. Equipment 100 square and a calculator.

i. Red goes first and is given a random whole number on the square. Say 65. Red multiplies 65 by something, say 0.25 giving 16.25.

Red colours the squares with X's as below.

They being the rounded 16.25 and its surrounding squares. Red has scored 9.

91	92	93	94	95	96	97	98	99	100
81	82	83	84	85	86	87	88	89	90
71	72	73	74	75	76	77	78	79	80
61	62	63	64	65	66	67	68	69	70
51	52	53	54	55	56	57	58	59	60
41	42	43	44	45	46	47	48	49	50
31	32	33	34	35	36	37	38	39	40
21	22	23	24	~~25~~	~~26~~	~~27~~	28	29	30
11	12	13	14	~~15~~	~~16~~	~~17~~	18	19	20
1	2	3	4	~~5~~	~~6~~	~~7~~	8	9	10

Fig. 1

ii. Blue takes the 16 as a starting number and multiplies it by 5 giving 80 marking the surrounding squares with O's. The board then looks like this:

The score is now Red: 9 Blue: 6

91	92	93	94	95	96	97	98	99	100
81	82	83	84	85	86	87	88	(89)	(90)
71	72	73	74	75	76	77	78	(79)	(80)
61	62	63	64	65	66	67	68	(69)	(70)
51	52	53	54	55	56	57	58	59	60
41	42	43	44	45	46	47	48	49	50
31	32	33	34	35	36	37	38	39	40
21	22	23	24	~~25~~	~~26~~	~~27~~	28	29	30
11	12	13	14	~~15~~	~~16~~	~~17~~	18	19	20
1	2	3	4	~~5~~	~~6~~	~~7~~	8	9	10

Fig. 2

iii. Red starting with 80 multiplies by 1.25 giving 100. The board then is:

The score is Red: 11 Blue: 4

Red has removed 2 from Blue by landing next to Blue. 89 and 90 have been neutralised but cannot be reclaimed.

91	92	93	94	95	96	97	98	~~99~~	~~100~~
81	82	83	84	85	86	87	88	~~(89)~~	~~(90)~~
71	72	73	74	75	76	77	78	(79)	(80)
61	62	63	64	65	66	67	68	(69)	(70)
51	52	53	54	55	56	57	58	59	60
41	42	43	44	45	46	47	48	49	50
31	32	33	34	35	36	37	38	39	40
21	22	23	24	~~25~~	~~26~~	~~27~~	28	29	30
11	12	13	14	~~15~~	~~16~~	~~17~~	18	19	20
1	2	3	4	~~5~~	~~6~~	~~7~~	8	9	10

Fig. 3

iv. If a number has already been taken or if off the board then that player loses their go.

v. The game ends when all the squares have gone, the winner being the one with most squares.

(The generation of the initial random number with a calculator, can provoke some interesting discussion. Alternatively a non participant can give one).

C. Variations

(a) The game can be played solo. The object then being the least number of goes.
(b) The size and shape of the board can be varied. A useful one is to have just a number line. (Then you claim your number and the one either side).
(c) Truncating results instead of rounding.
(d) $+$ $-$ and $\div$ can be used instead of $\times$
(e) Firing can be from a fixed position instead of a moving one. e.g. Always do $100 - \square =$ to move.
(f) "Unary" functions (the ones on the calculator that alter one number such as $\sqrt{}$, x^2, 10^x) can be used as 'guns' (pardon the regrettable but so natural military connotations) e.g. aiming at 40 might entail 1600 $\sqrt{}$.
(g) Combinations of operations, say, enter x do $x \times 3 + 10$ and that gives the number.

D. If the reader is bewildered by the variations then (s)he joins the author. There seems to be such a rich source of games here, so easily tailored to particular needs that its hard to know where to begin.

I suppose the thing to do is adapt it to suit the thing you are currently working on. e.g. Suppose you had been working on operations with negative numbers: The shape could be a line. The numbers changed to -2 to $+2$, regions could be 0.1 either side of the integers and the rest:

Below [−2] [−1] [0] [1] [2] Above

(11 regions including ABOVE and BELOW) Multiplying and adding allowed. Move from where you last landed. Only get the region you land on.

E. Another game from the same family that I learnt from Ken Tyler is this: the first player enters a number. The other multiplies it by something trying to get it in [100,101]. If not successful the other player tries by multiplying from the current number in the display. It has proved very popular with 11-14 year olds.

F. I showed my middle ability 14 years olds the game during a lesson, offering them duplicated 100 squares. I explained the rules and showed them 2 such as $80 \times 0.5 = 40$ and $30 \times 1.76 = 52.8$ and let them get on, encouraging them to guess when they'd got a rough idea. Some needed help on various numbers. The main advice was "What happens when you multiply by 1, $\frac{1}{2}$, or $1\frac{1}{2}$?"

I left them to play it in their "form periods" and on the one lesson a week I set aside for work not directly connected with $\frac{1}{2}$ termly tests (we were on a very tight rein).

Observing the mathematical strategies adopted by the children showed two main approaches.

The first depended on the distribution of multiplication over addition and went roughly $73 \times ? = 46$.

(a) What $73 \times ? = \square$ bonds do I know?
(b) $73 \times 0.5 = 36.5$; $73 \times 0.1 = 7.3$ and $73 \times 0.01 = 0.73$
(c) Make up to 46 using these so

$$36.5 + 7.3 + 0.73 + 0.73 + 0.73 = 45.99$$

so

$$73 + (0.5 + 0.1 + 0.01 + 0.01 + 0.01) = 73 \times 0.63 = 45.99$$

It was rarely done as thoroughly as this.

A variation was to find out what to take off:

$$73 - 46 = 27 \text{ so } 73 \times (1 - ?) = 46$$

$$73 \times 0.37 = 27 \text{ roughly so } 73\ (1 - 0.37) = 46$$

A problem with this approach is that you need a very good memory. They were loathe to use paper as a temporary store.

The way they used it had many variations but this was typical.

$$73 \times ? = 46.$$

$$73 \times 0.5 = 36.5$$

$$73 \times 0.6 = 43.8\ (36.5 + 7.3) \text{ too low}$$

$$73 \times 0.7 = 51.1\ (43.8 + 7.3) \text{ too high}$$

$$73 \times 0.6 = 43.8$$

$$73 \times 0.61 = 44.5 \text{ roughly } (43.8 + 0.7)$$

$$73 \times 0.62 = 45.2 \text{ roughly}$$

$73 \times 0.63 = 45.9$ roughly — Hope that the neglected bits don't make the 45.9 go over 46.5 ... so use 0.63.

The other main way was to use fractions. This dependend on their knowing their tables and the calculator decimal equivalent of fractions. It wasn't used as much as I'd hoped it would be.

$36 \times ? = 27$	Thinks:both in $9 \times$ table

so $36 \div 4 \times 3 = 27$ so $36 \times 3/4 = 27$ use *0.75*

It is surprising how many can be done in a similar way with a bit of fudging, e.g. the $73 \times ? = 46$. It's almost $72 \times ? = 45$ and $72 \div 8 \times 5 = 45$ so $72 \times 5/8 = 45$ ... so try 5/8.

Few used the above ways to get precise answers. They did as much as they could and guessed the final bits of the multiplier.

I encouraged this guessing, feeling that the skills underlying good guessing give a view of whats happening, uncluttered by computation. To help this I suggested being allowed two goes at a number so the first could be a rough guess to be refined, but this wasn't well received as bad first guesses caused players to claim they were aiming elsewhere. (I still think its a good idea. Extending it, a player could keep trying until successful and the number of goes needed accumulated to work out the winner.)

About four weeks later I discussed how they had tackled various numbers and suggested that the approach of working out $1\times$, $0.1\times$, and $0.01\times$ of a number and then making up the required ones from these bits would work for all numbers. I gave them some to practice while emphasising they should use other ways when they seemed simpler.

Most enjoyed the game but this final drilling took the edge off it and I would avoid it another time. I didn't make the task easier for those that found it hard and this could easily have been done by using a number line with 1 to 20 on — then they wouldn't have had to consider any multipliers but those with 0 or 1 decimal place of 2 decimal places with a 5 in the hundredths column.

G. The games drawn in B and D are written as the computer programs, NCROACH and LINEOUT on the "Teaching with a Micro: Maths 4" package, along with three other programs and published by ITMA, Shell Centre, Nottingham University.

They are just as shown here, the advantage of a computer being it does the administration of the game which helps, particularly in NCROACH, but it is certainly not essential.

For further ideas we suggest you refer to the Open University's course booklet "Calculators in Primary Schools", The Nuffield 5-11 "Electronic Calculators" Teachers' Handbook and Walsall Primary Schools' Calculator Guide.

FRACTIONS BY CALCULATOR

by William Wynne Willson, Faculty of Education, University of Birmingham

Several recent issues of *Mathematics in School* have included interesting and provocative points about the place that fractions should or should not have in the curriculum (see D. G. H. Lloyd, "The Case Against Fractions", Vol. **5**, No. 2, March 1976; R. R. Joy, "In defence of Fractions", Vol. **5**, No. 4, September 1976; John Cable, "A society for the preservation of fractions", Vol. **5**, No. 4, September 1976; W. W. Sawyer, "The Case for Fractions", Vol. **6**, No. 1, January 1977): The use of calculators in the classroom ought nowadays to be a significant ingredient of this discussion. Any calculator makes it easy to cope with decimals; at present only a few calculators (such as the Casio fx 120) are designed to do calculations in fractions, and since this facility is seldom required in applications outside school it seems unlikely to become a common feature of other machines. The purpose of this article is to describe a straightforward process for working in fractions on any calculator. The method is reasonably easy and natural, so that it could provide a way of introducing fractions to children brought up on calculators, or even an alternative to learning how to manipulate them: such issues are not discussed here.

The problem

With a calculator available, it is easy enough to simplify expressions involving fractions: one simply treats $\frac{3}{5}$ as a way of writing $3 \div 5$. Thus for example:

$$\frac{3}{5}-\frac{1}{2} = 0.6-0.5 = 0.1,$$
$$\frac{1}{4}+\frac{3}{10} = 0.25+0.3 = 0.55,$$
$$\frac{\frac{3}{5}}{\frac{5}{16}} = 0.6 \div 0.312\,5 = 1.92$$

or even

$$1\tfrac{1}{17}+1\tfrac{1}{4} = 1.058\,823\,5\ldots+1.25$$
$$= 2.308\,823\,5\ldots$$

But it may be objected that the answers are unsatisfactory as they are not in fraction form. The required answers are

$$\tfrac{1}{10},$$
$$\tfrac{11}{20} \text{ (not } \tfrac{55}{100} \text{ nor } \tfrac{22}{40}\text{)},$$
$$1\tfrac{23}{25} \text{ (or perhaps } \tfrac{48}{25}\text{)},$$
$$\text{and } 2\tfrac{21}{68} \text{ respectively.}$$

Thus the *essential problem is to convert a decimal* (expressed to a limited number of figures) *back into a fraction in its lowest terms*. It is remarkable that there is a simple way to do this, which as we shall see achieves a good deal more as well.

The next three sections of this article provide an introduction to the method which attempts also to explain how it is derived and how it works. If you only want to use the method you can safely skip much of the explanation; but you will need to master the flow diagram in the general method and the table in the second stage.

The first stage

Consider first the simple problem of fractions of the form $\frac{1}{x}$. You may not recognise

$$0.034\,482\,75\ldots$$

but if you are told that it is the reciprocal of a whole number it is obvious that by dividing it into 1 (or using the reciprocal key if there is one on your calculator) the value of x can be found. In fact

$$\frac{1}{0.034\,482\,75\ldots} = 29$$

There are problems here about "rounding-off", but this is not a practical difficulty. Thus if the decimal is entered as 0.034 482 75, its reciprocal appears as 29.000 007 25 . . . We take it for granted that this "ought to be" 29, and will continue to take similar liberties throughout the rest of this article. If you are more punctilious you may like to check that

$$\frac{1}{0.034\ 482\ 75} > 29 > \frac{1}{0.034\ 482\ 76}, \text{ thus}$$

completely justifying the use of the equality sign above.

Taking the reciprocal is helpful even when the number is *not* of the form $\frac{1}{x}$. For example to find a fraction equal to 0.307 692 3 . . . note that

$$\frac{1}{0.307\ 692\ 3} = 3.250\ 000\ 0\ldots$$

which is expressible as $3\frac{1}{4}$. Hence

$$0.307\ 692\ 3\ldots = \frac{1}{3\frac{1}{4}} = \frac{1}{\frac{13}{4}} = \frac{4}{13}$$

This process does not really depend on recognising 0.25 as $\frac{1}{4}$, since if necessary we could have done that conversion by finding $\frac{1}{0{\cdot}25} = 4$. As another illustration of this two-stage process, consider 0.490 566 03 . . .

$$\frac{1}{0.490\ 566\ 03\ldots} = 2.038\ 461\ 5\ldots$$

Do you recognise 0.038 461 5 . . .? Probably not, so try finding its reciprocal:

$$\frac{1}{0.038\ 461\ 5\ldots} = 26$$

Hence $0.490\ 566\ 03\ldots = \dfrac{1}{2+\frac{1}{26}} = \dfrac{26}{53}$.

If you would like to check your grasp of this idea, try using the method to express

0.325
and 0.454 545 45 . . .

as fractions. (You can of course check your answers by division.)

The general method

You may well be thinking that this approach is rather limited since it only applies to particular fractions, namely those of the form $\frac{1}{x}$ or $\cfrac{1}{x+\cfrac{1}{y}}$

But the point is that *any* fraction can be expressed in the form

$$x+\cfrac{1}{y+\cfrac{1}{z+\cfrac{1}{\ddots\ \ldots+\cfrac{1}{w}}}}$$

so that if you are prepared to go on rather longer the method will work with any rational number.

Here is a flow diagram describing how to perform the process, with the aid of a calculator and pencil and paper.

1. Set the given number on the calculator.
2. Note the integer part on paper.
3. Is the decimal part zero?* If so, stop. Otherwise continue.
4. Subtract the integer part.
5. Take the reciprocal.
6. Go back to 2.

As an illustration here is the working for 3.141 592 92.

* As mentioned above, common sense about ''rounding-off'' is need: for example with 3.000 003 21 . . . in the display you should probably answer ''Yes''.

Step		Calculator	Paper
1		3.141 592 92	
2		3.141 592 92	3
3	No		
4		0.141 592 92	3
5		7.062 500 02	3
2		7.062 500 02	3, 7
3	No		
4		0.062 500 02	3, 7
5		15.999 995 48	3, 7
2		15.999 995 48	3, 7, 15
3	No		
4		0.999 995 48	3, 7, 15
5		1.000 004 52	3, 7, 15
2		1.000 004 52	3, 7, 15, 1
3	Yes (near enough!)		

This shows that $3.141\ 592\ 92 \approx 3+\cfrac{1}{7+\cfrac{1}{15+\cfrac{1}{1}}}$

$$= 3+\frac{1}{7\frac{1}{16}} = 3\frac{16}{113} = \frac{355}{113}$$

Note that if you had assumed (very reasonably) that 15.999 995 48 ought to be 16.000 . . . you would have stopped a stage earlier with 3, 7, 16 on your paper, which would have given the same result.

A valuable feature of this method is that it always gives the answer in its lowest terms. For example consider

$$\frac{1760}{1728} = 1.018\ 518\ 5$$

Applying the method to 1.018 518 5 gives 1, 54 so that

$$\frac{1760}{1728} = 1+\frac{1}{54} = \frac{55}{54}$$

As a third worked example consider 0.477 124 18 . . . This should be found to give 0, 2, 10, 2, 3 and thus

$$0.477\ 124\ 18 = 0+\cfrac{1}{2+\cfrac{1}{10+\cfrac{1}{2+\cfrac{1}{3}}}} = 0+\cfrac{1}{2+\cfrac{1}{10+\cfrac{3}{7}}}$$

$$= 0+\cfrac{1}{2+\cfrac{7}{73}}$$

$$= \frac{73}{153}$$

The second stage

A valid criticism of this approach as described so far is that it can be laborious to sort out the fraction after the ''multiple-decker'' form has been obtained, and that this part of the work in itself demands a good understanding of fractions.

Luckily this calculation can be done more easily by a process which starts (unexpectedly) from the *top* of the fraction rather than the bottom. Consider again the last example for which the flow diagram gave 0, 2, 10, 2, 3. The new process provides in turn the values of

$$0,\ 0+\tfrac{1}{2},\ 0+\cfrac{1}{2+\cfrac{1}{10}},\ 0+\cfrac{1}{2+\cfrac{1}{10+\cfrac{1}{2}}} \text{ and } 0+\cfrac{1}{2+\cfrac{1}{10+\cfrac{1}{2+\cfrac{1}{3}}}}$$

that is

$0, \frac{1}{2}, \frac{10}{21}, \frac{21}{44}$ and $\frac{73}{153}$ respectively.

It is convenient to arrange the work in a table.

The top row is the "print-out" from the flow diagram; the second row, giving the numerators of the fractions starts with 0, 1; the third row which gives the denominators starts with 1, 0.

Print-out from flow diagram			0	2	10	2	3
Numerators	0	1					
Denominators	1	0					
Fractions	$\frac{0}{1}$	$\frac{1}{0}$					

The numerator and denominator rows are completed separately from left to right. At each stage the next value is found by multiplying the number at the head of the column by the value to the left of the one you want and then adding the value to the left of that. Thus in the example above the successive entries for the numerator row are:

		0	2	10	2	3
0	1	$0\times1+0$ $=0$	$2\times0+1$ $=1$	$10\times1+0$ $=10$	$2\times10+1$ $=21$	$3\times21+10$ $=73$

Similarly the denominators are:

1	0	$0\times0+1$ $=1$	$2\times1+0$ $=2$	$10\times2+1$ $=21$	$2\times21+2$ $=44$	$3\times44+21$ $=153$

Thus the complete calculation in this case gives:

Print out			0	2	10	2	2
Numerators	0	1	0	1	10	21	73
Denominators	1	0	1	2	21	44	153
Fractions	$\frac{0}{1}$	$\frac{1}{0}$	$\frac{0}{1}$	$\frac{1}{2}$	$\frac{10}{21}$	$\frac{21}{44}$	$\frac{73}{153}$

Although this takes some time to describe it is quite quick to do.

Moreover it can, if you like, be done at the same time as the original process. In fact if your calculator has enough memories (five are needed) the whole thing can be done without writing anything down.

As another example consider 3.141 592 92 which gave the print-out 3, 7, 15, 1, the table will be:

		3	7	15	1
0	1	3	22	333	355
1	0	1	7	106	113
$\frac{0}{1}$	$\frac{1}{0}$	$\frac{3}{1}$	$\frac{22}{7}$	$\frac{333}{106}$	$\frac{355}{113}$

It is not too difficult to justify this process by mathematical induction. (See any book on Continued Fractions, e.g. ref. 1.)

Approximations

There is a pleasant extra advantage in calculating the fraction by the method described in the last section: if we look at the fractions generated we find that they form a sequence which rapidly approaches the final value. This can be checked by converting them back into decimals. For example if we apply the process 3.145 926 5 (which is the value of π correct to 8 places of decimals) the table produces :

		3	7	15	1	288	1	2
$\frac{0}{1}$	$\frac{1}{0}$	$\frac{3}{7}$	$\frac{22}{7}$	$\frac{333}{106}$	$\frac{355}{113}$	$\frac{102\,573}{32\,650}$	$\frac{102\,928}{32\,763}$	$\frac{308\,429}{98\,176}$

1	3	1	7	3	1
$\frac{411\,357}{130\,939}$	$\frac{1\,542\,500}{490\,993}$	$\frac{1\,953\,857}{621\,932}$	$\frac{15\,219\,499}{4\,844\,517}$	$\frac{47\,612\,354}{15\,155\,483}$	$\frac{62\,831\,853}{20\,000\,000}$

To eight places we have

$\frac{3}{1}=3.00\,000\,000$
$\frac{22}{7}=3.142\,857\,14\ldots$
$\frac{333}{106}=3.141\,509\,43\ldots$
$\frac{355}{113}=3.141\,592\,92\ldots$
$\frac{102\,573}{32\,650}=3.141\,592\,64\ldots$
$\frac{102\,928}{32\,763}=3.141\,592\,65\ldots$

All the remaining fractions give 3.141 592 65 to eight places. The process has provided a series of increasingly accurate approximations for π, including the well-known $\frac{355}{113}$ (see ref. 2 for good comments on $\pi\approx3\frac{1}{7\frac{1}{16}}$).

Similarly the method can be used to find simple fractions which approximate to any number. For example the number of days in a year is 365.242 195. . . . Now .242 195 . . . gives us:

		0	4	7	1	3	. . .
$\frac{0}{1}$	$\frac{1}{0}$	$\frac{0}{1}$	$\frac{1}{4}$	$\frac{7}{29}$	$\frac{8}{33}$	$\frac{31}{128}$	. . .
		0	0.25	0.2413 . .	0.2424 . .	0.242 187 5	. . .

Julius Caesar's calendar, with 1 leap year in 4 years, corresponds to $\frac{1}{4}$. An arrangement of the calendar with 8 leap years in every 33 is said to have been adopted in twelfth century Persia on the advice of Omar Khayyám: it is in fact more accurate than our present arrangement (Pope Gregory's plan with 97 leap years in 400 years) because $\frac{97}{400}=.2425$. A calendar with 31 leap years in 128 would give an error of less than 1 day in 10 000 years.

Again, if we apply the process to e$=2.718\,281\,828\,45\ldots$ we get 2, 1, 2, 1, 1, 4, 1, 1, 6, 1, 1, 8, 1, 1, 10, 1, 1 . . . giving

$\frac{2}{1}, \frac{3}{1}, \frac{8}{3}, \frac{11}{4}, \frac{19}{7}, \frac{87}{32}, \frac{106}{39}, \frac{193}{71},$

$\frac{1264}{465}, \frac{1457}{536}, \frac{2721}{1001}\ldots$

Here $\frac{193}{71}=2.718\,30\ldots$ is correct to four places of decimals so that $2\frac{51}{71}$ is a good approximation to e, while $\frac{2721}{1001}$ has an error of only $-0.000\,000\,11\ldots$

Conclusion

This method of handling fractions on a calculator is easy to use with a little practice. It also provides a convenient way to reduce a fraction to its lowest terms (not depending on primes), and to provide fractions which approximate to given decimals. Finally it may be noted that if you have a programmable calculator or computer terminal available it is an interesting exercise to write a programme to operate the whole process.

References

1. Olds, C. D. (1963) *Continued Fractions*. New Mathematical Library.
2. Hope-Jones, W. (1971) "Simplicity and truthfulness in Arithmetic". *Mathematical Gazette*, Vol. XIII, No. 253 (1939). Reprinted in Vol. LV, No. 392 (1971).

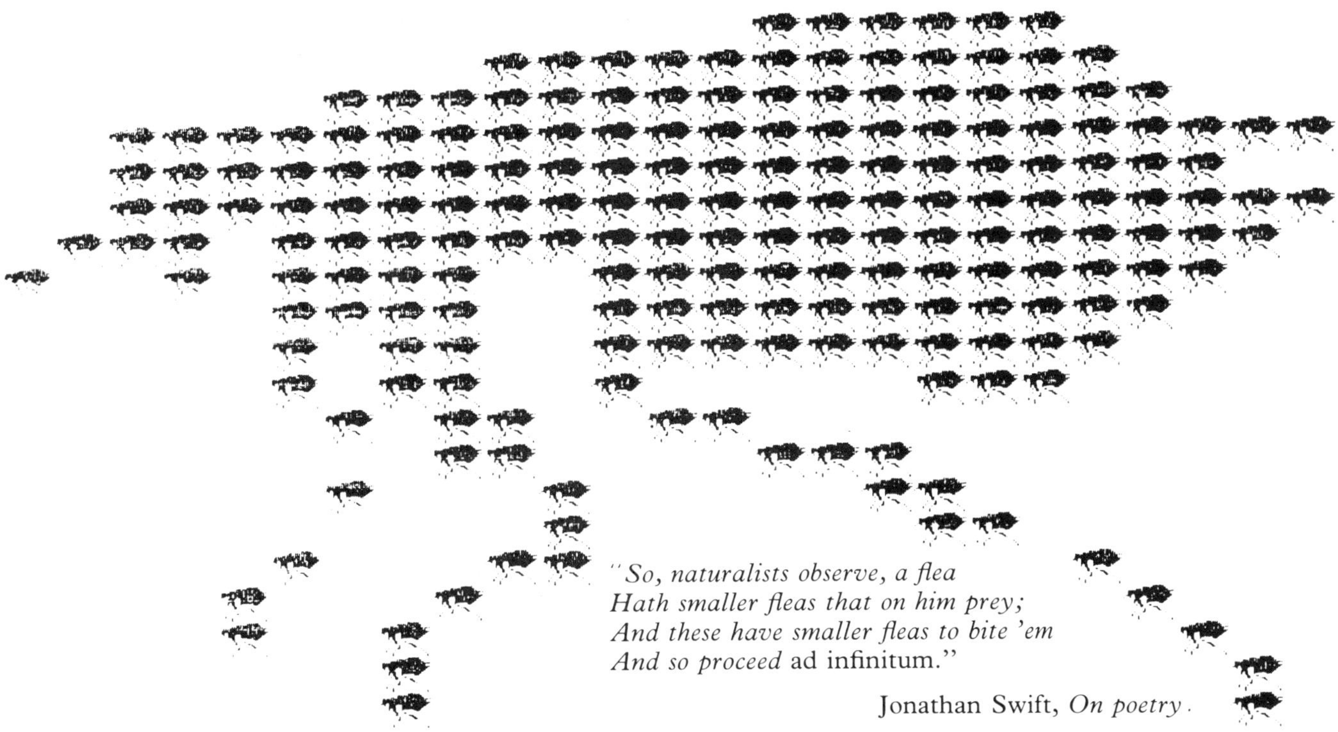

"So, naturalists observe, a flea
Hath smaller fleas that on him prey;
And these have smaller fleas to bite 'em
And so proceed ad infinitum."

Jonathan Swift, *On poetry*.

Fractal Patterns on a Square

by John K. Backhouse
Department of Educational Studies,
University of Oxford

A good many years ago my imagination was caught by figures illustrating the "snowflake" curve in *Mathematics and the imagination* by Kasner and Newman[1]. So it was not surprising that an article by Richard Noss should revive it[2]. In that article there was a description of the "Koch curve" which is shown in Figures 1a to d.

Look at the *first order* curve in Figure 1a. It may be described by the following instructions on the left

draw a line	draw the curve of next order below
turn left	turn left
draw a line	draw the curve of next order below
turn right	turn right
draw a line	draw the curve of next order below
turn right	turn right
draw a line	draw the curve of next order below
turn left	turn left
draw a line	draw the curve of next order below

Higher order curves may then be generated by the instructions on the right. The second order curve consists of five first order curves with turns of left, right, right and left

Fig. 1a

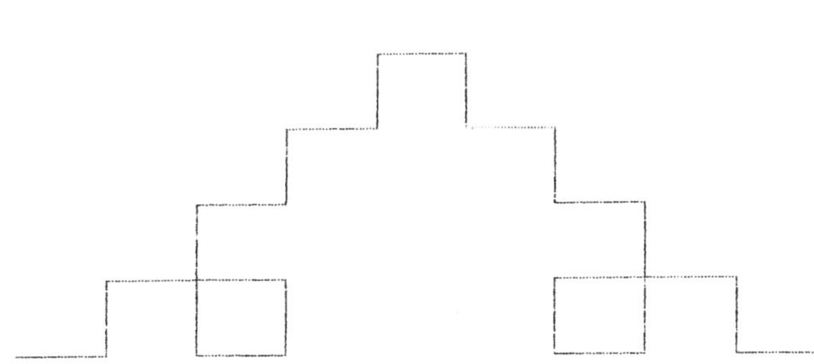

Fig. 1b

between them, as may be seen in Figure 1b. Then third and fourth order curves are shown in Figures 1c and 1d.

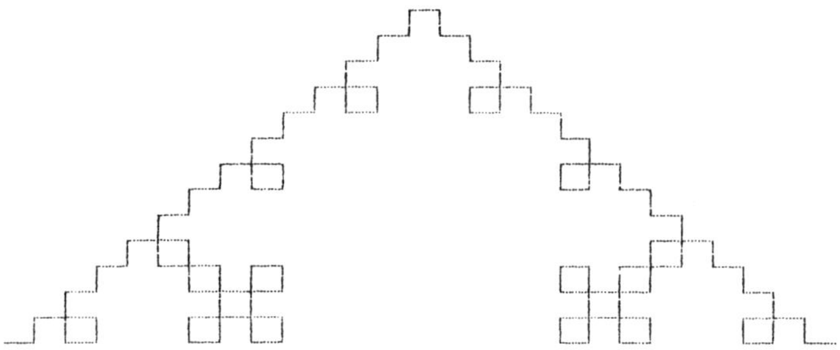

Fig. 1c

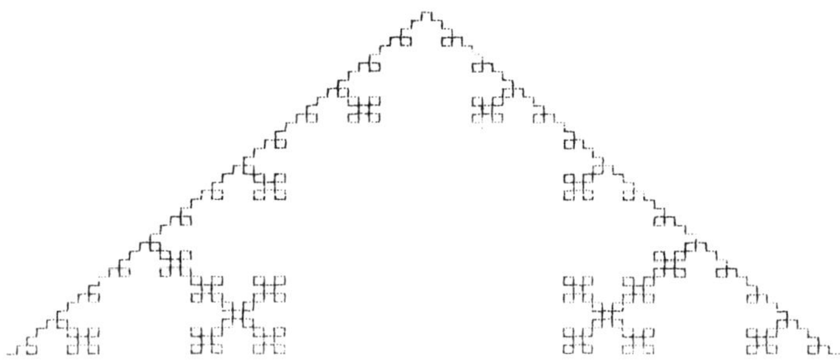

Fig. 1d

The title page of Richard Noss's article includes a picture of the "Menger sponge" which is in three-dimensional space. (I shall return to the sponge soon.) However he appears to make no mention of what might be called "plane fractals". This article is directed at these and a programme written in BBC BASIC is appended so as to encourage readers who are loth to write their own programmes to make their own fractal-based patterns. I hope that it may encourage a creative activity within a mathematical context. Like the "132 short programmes" on the Mathematical Association disc, this programme is intended to be modified by the user and, as will be seen, it presents a considerable range of possibilities. The figure at the top of the article shows a fanciful illustration of my quotation from Swift. Each "flea" is made up of 207 white pixels on the screen and the whole figure is produced by using "fleas" instead of dots. Travellers on the Bakerloo Line may have noticed tile patterns depicting Sherlock Holmes constructed on similar lines. Figure 2 shows a much simpler possibility. The basis of the design, a *first order* pattern, is a 3 by 3 square

0	1	1
1	0	1
1	1	0

where "0" and "1" denote pixels on the screen coloured black and white respectively, while on the printed paper the colours are reversed. (Sorry about it but this is the way the screen dump which produced the figures works.) The pattern matrix appears in lines 600 to 620 of the programme at the end of this article. The second order figure follows the same pattern but now the "0" and "1" represent a black 3 by 3 square and a first-order pattern respectively on the screen, the colours again being reversed on paper. (This is analagous to the situation described in the first paragraph where a line in the basic design if replaced by the curve of the next order below when drawing curves of higher order.)

In Figure 2 it may be seen how this process is continued up to the fifth order pattern.

Fig. 2

The largest square which can be drawn on the BBC micro is 256 by 256 pixels. (The "flea" would not fit on to the screen and was constructed in six parts.) This means that a first order pattern on a 2 by 2 square can generate an eighth order pattern, as in Figure 3. The data matrix was

```
6ØØ DATA 1,1
61Ø DATA Ø,1
```

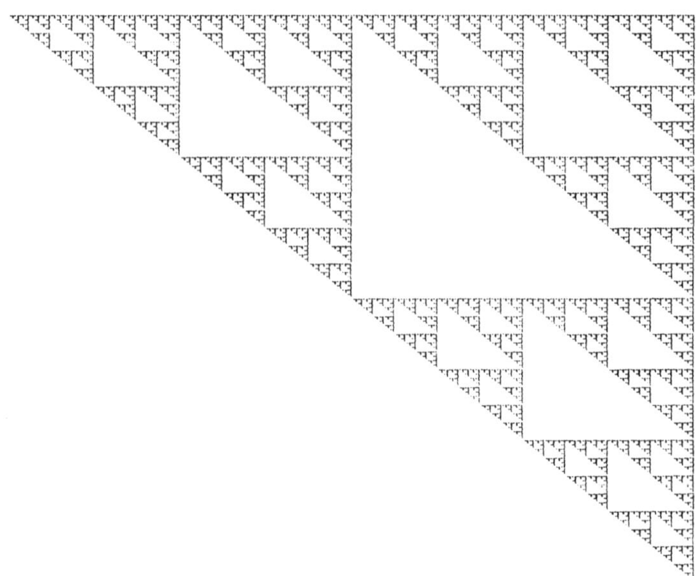

Fig. 3

There is such limited scope for generating patterns starting from a 2 by 2 square that the programme does not cover this case. Starting with a 3 by 3 square, the highest order that can get on the screen is the fifth with 243 by 243 pixels. Such are the dimensions of this square that those familiar with the Cockcroft Report[3] may well ask "Is there an investigation here?" The answer is that there is plenty of scope for the question "What happens if ...?" Some illustrations of the varied patterns that may be obtained now follow. Figure 4 was obtained from a 6 by 6 data matrix

```
600 DATA 0,0,0,1,0,0
610 DATA 0,1,0,1,1,0
620 DATA 1,1,1,1,0,0
630 DATA 0,0,1,1,1,1
640 DATA 0,1,1,0,1,0
650 DATA 0,0,1,0,0,0
```

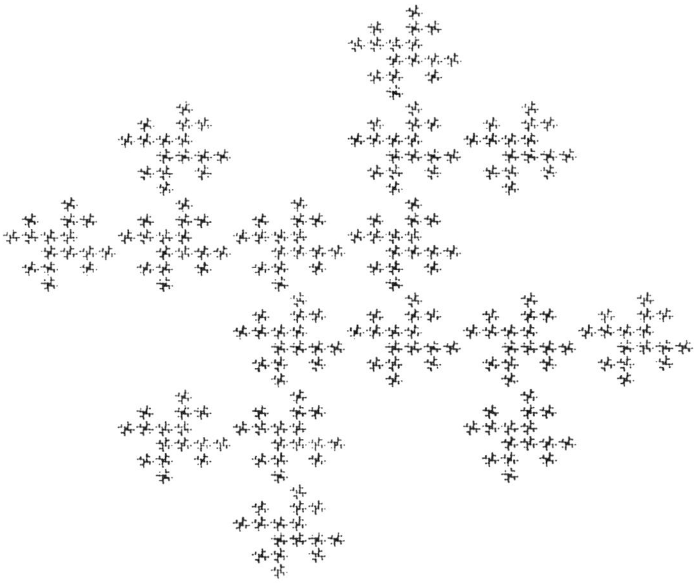

Fig. 4

Figure 5 is based on a 3 by 3 data matrix

```
600 DATA 0,1,0
610 DATA 1,1,1
620 DATA 0,1,0
```

If you turn the figure through about 45 degrees, you may see the outline of Figure 1d.

Fig. 5

Then Figure 6 is based on a 3 by 3 data matrix

```
600 DATA 1,1,1
610 DATA 1,0,1
620 DATA 1,1,1
```

This brings us back to Menger's sponge because the pattern represents one face of it. Imagine that twenty Centicubes are stuck together so that they make the outline

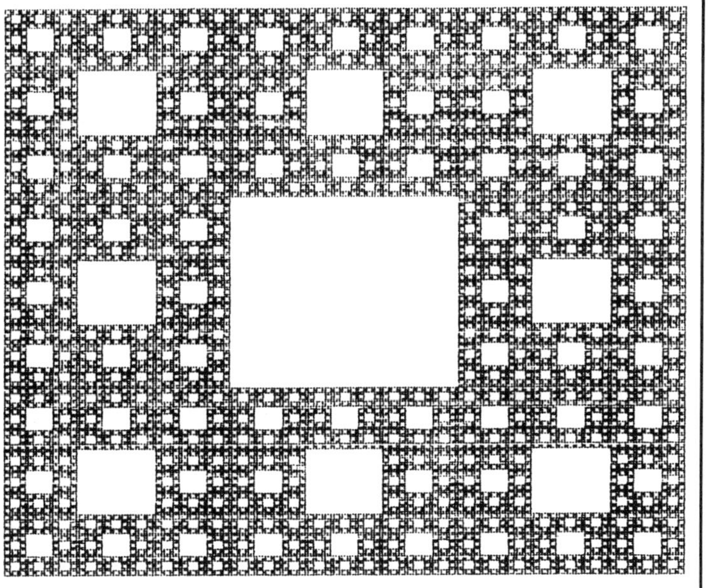

Fig. 6

of a 3 cm by 3 cm cube with holes passing through the centre from one face to the opposite one. This forms the first order pattern for the Menger sponge.

Changing the rules

Patterns produced by the programme as it stands may be considered rather austere, but they may be modified in a number of ways, some of which are illustrated below. Many people enjoy tessellations and Figure 7 is a modification of the pattern in Figure 2. Instead of producing a fifth order pattern a tessellation of the fourth order one has been printed out. To do this, simply delete line 120 from the programme.

Fig. 7

Changing the action taken at the highest order is something which can be emulated at lower orders. Figure 8 is a third order fractal pattern, the basis for the design being the data matrix

```
600 DATA 1,0,1,0,1
610 DATA 0,1,0,1,0
620 DATA 1,0,0,0,1
630 DATA 0,1,0,1,0
640 DATA 1,0,1,0,1
```

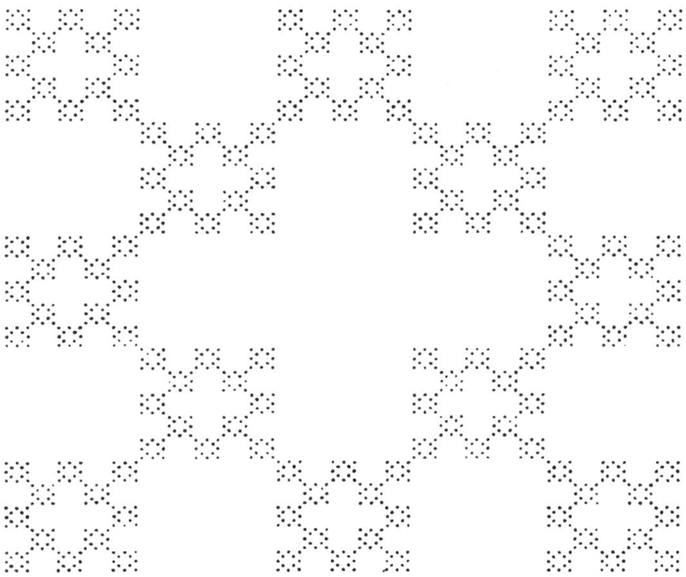

Fig. 8

However the programme may be modified to produce Figure 9 by changing line 230 to

```
230 PROCCHANGE(X2,Y2): IF SW=0 THEN GOTO 330
```

and writing a new procedure

```
800 DEF PROCCHANGE(X,Y)
810 SW=1
820 IF (X=0 OR X=4) AND Y MOD 2=1 THEN
    SW=0
830 IF X=1 OR X=3 THEN SW=0
840 IF X=2 AND (Y>0 AND Y<4) THEN SW=0
850 ENDPROC
```

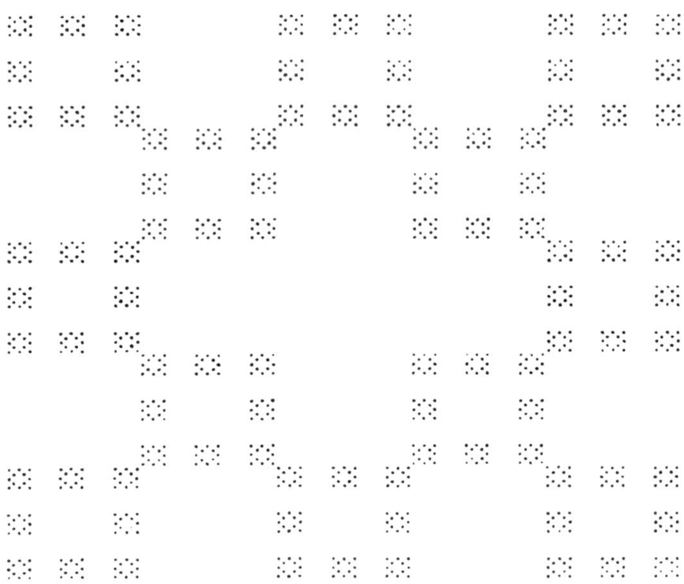

Fig. 9

This is probably simpler than inserting another data matrix and reading it in. When a first order pattern is wanted on the screen, SW is given the value 1, otherwise SW = 0.

Figure 10 is a modification of Figure 2 in which the third order pattern is composed of four second order patterns in their normal orientation together with two which have been reflected about a line parallel to the edge of the page. This result may be achieved by inserting a line

```
285 IF Y3=1 THEN YP=YP+(8-2*(Y1+3*Y2))*K
```

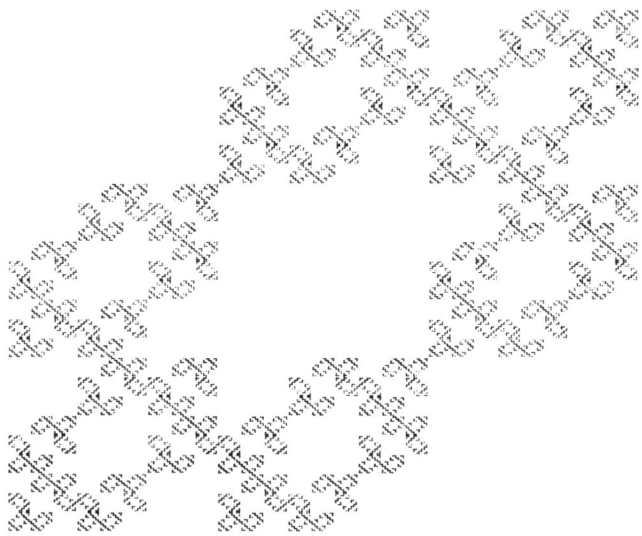

Fig. 10

and Figure 11 has been obtained by the insertion

```
285 IF (X3+Y3=1 OR X3+Y3=4) THEN
    YP=YP+(8-2*(Y1+3*Y2))*K
```

Fig. 11

The reader may like to see what happens if, in addition, the fourth order pattern is tessellated instead of producing a fifth order pattern after the usual manner. Once the idea of using transformations has been suggested, there is tremendous scope for investigating the possibilities of pattern making. It is hoped that readers will develop the ideas further – one obvious omission is the use of colour.

References

1. Kasner, E. and Newman, J. (1949) *Mathematics and the Imagination*, London: G. Bell and Sons Ltd.
2. Noss, R. (1985) "Fractal, Turtles and Snowflakes", *Micromath* **1**(1), 30–35.
3. DES, (1982) *Mathematics Counts*. Report of the Committee of Inquiry into the Teaching of Mathematics in Schools. London: HMSO.

```
10 REM: FRACMAT
20 MODE 1: VDU28, 34,31,39,0
30 X5=0: Y5=0: X4=0: Y4=0: X3=0: Y3=0:
   X2=0: Y2=0
40 INPUT "what order of fractal"; F
50 INPUT "What dimension of square";
   N: N=N-1
60 DIM MATRIX(16.16): GOSUB 500
70 K=4: H=N+1
80 ON F GOTO 240, 190, 160, 130, 100
100 FOR X5=0 TO N
110 FOR Y5=0 TO N
120 PROCHOLES(X5,Y5): IF SW=0 THEN GOTO
    420
130 FOR X4=0 TO N
140 FOR Y4=0 TO N
150 PROCHOLES (X4,Y4): IF SW=0 THEN GOTO
    390
160 FOR X3=0 TO N
170 FOR Y3=0 TO N
180 PROCHOLES(X3,Y3): IF SW=0 THEN GOTO
    360
190 FOR X2=0 TO N
200 FOR Y=0 TO N
230 PROCHOLES(X2,Y2): IF SW=0 THEN GOTO
    330
240 FOR X1=0 TO N
250 FOR Y1=0 TO N
260 PROCHOLES(X1,Y1): IF SW=0 THEN GOTO
    300
270 XP=(X1+X2*H+X3*H*H+X4*H*H*H
        +X5*H*H*H*H)*K
280 YP=(Y1+Y2*H+Y3*H*H+Y4*H*H*H
        +Y5*H*H*H*H)*K
290 PLOT69, XP, YP
300 NEXT Y1
310 NEXT X1
320 IF F=1 THEN STOP
330 NEXT Y2
340 NEXT X2
350 IF F=2 THEN STOP
360 NEXT Y3
370 NEXT X3
380 IF F=3 THEN STOP
390 NEXT Y4
400 NEXT X4
410 IF F=4 THEN STOP
420 NEXT Y5
430 NEXT X5
440 STOP
500 REM INPUT OF MATRIX
510 FOR Y=N TO 0 STEP -1
520 FOR X=0 TO N
530 READ MATRIX (X,Y)
540 NEXT X
550 NEXT Y
560 RETURN
600 DATA 0,1,1
610 DATA 1,0,1
620 DATA 1,1,0
700 DEF PROCHOLES(X,Y)
710 SW=1
720 IF MATRIX(X,Y)=0 THEN SW=0
730 ENDPROC
```

CALCULATOR CORNER

A series of games and puzzles for which a calculator could be useful. They have been submitted by Barry Blakeley on behalf of the calculator sub-committee of the MA Teaching Committee.

Unarmed combat

For two players, one calculator, and a set of "problem" cards.

On one side of each card there is an arithmetic problem suitable for the ability of the players (e.g. 18 − 6, (2 × 5) + 2, 36 ÷ 2). Pupils may even decide upon these themselves.

One player, A, uses the calculator.

The other player, B, is "unarmed".

B turns over a card and tries to work out the answer before A announces the result from the calculator. The first player to produce the correct result wins the point.

After five turns the players change over so that B uses the calculator and A turns over the cards.

The winner is the first player to get 10 points.

Bull's Eye

For any number of players, one calculator per player, pencil and paper.

Choose a two-digit number as the target and four digits as the ammunition.

Try to hit the target using only the specified ammunition (not repeated) and the operations +, −, ×, ÷.

Score
10 for a direct hit
7 for an "inner" (within 2 of target)
3 for an "outer" (within 5 of target)

Example
Target 58
Ammunition 4, 9, 3, 7

A: $4\times9+3\times7=57$ score 7
B: $(9+3)\times4+7=55$ score 3
C: $9\times7-(4+3)=56$ score 7
D: $7\times(4+3)+9=58$ score 10

Beware of calculator logic [4] [×] [9] [+] [3] [×] [7]

Make a more difficult game by:

Making the target a four-digit number and the ammunition four two-digit numbers.

Score 10 for a hit
7 for an inner (target ± 10)
3 for an outer (target ± 25)

5 The History of Mathematics
by Bud Winteridge

The use of historical material in the teaching of mathematics can have three positive effects.

First, it may enhance understanding of the subject. Often the stages through which a learner progresses follow closely the historical problems encountered by mathematicians in the past, and the way they were resolved, may help the learner to grasp the essential concepts of the topic.

Secondly, it may change the pupil's attitude to mathematics. A study of history suggests that mathematics is not a fixed body of knowledge handed down from above but it is a changing and developing subject. Mathematics is seen as being about processes rather than formalised blocks of knowledge.

Thirdly, it may improve motivation. It presents mathematics as being the outcome of human endeavour — something which has evolved and has been created as the result of considerable effort. Thus a historical approach will certainly increase the interest in the subject for some pupils.

For all these possible gains, there is little evidence of using aspects of the history of mathematics in mathematics lessons. Why is this? One possible reason is that many teachers themselves lack a knowledge of the history of mathematics. Over the years *Mathematics in School* has attempted to help teachers to gain some knowledge of the subject through various articles. *Historical Topics* by David Green, Roger Knott and Arthur Dodd were popular features of the journal. An example is included in this selection.

However there are dangers. As David Green pointed out in an article in 1976 (Vol 5 No 3) 'Popularisers have found the myths more tempting than the facts', with the result that often inaccuracies and misconceptions arise. Frank Tapson's article makes amusing reading but highlights the problems that arise from attempting to present complex ideas in a simple way to pupils with a limited mathematical background. Keith Lloyd (in Vol 6 No 5, 1977) raised the further problem of involving even undergraduates in anything more than copying material from standard texts on the history of mathematics because of the difficulty of obtaining and reading primary source material. However, there can be considerable benefits in using historical material in the teaching of mathematics provided the dangers and the limitations are recognised and provided the teacher ensures that accuracy rather than myth predominates.

Mathematics of ancient Babylon

by J. M. Dubbey, Polytechnic of the South Bank

Our knowledge of the mathematical achievement of any ancient civilization is dependent on a triple coincidence: that some form of record will be left, that these records will be discovered centuries later and that any mathematical content will be recognized and interpreted. In the case of the Babylonians it is fortunate that all these conditions have been satisfied. Their method of cuneiform writing consisting of wedge-shaped characters with a stylus on clay which was then baked out, ensured that a record of their civilization could be preserved. About half a million of these clay tablets have been found and the work of deciphering began early in the nineteenth century. More recently mathematical scholars, particularly Neugebauer and Thureau-Dangin, have investigated the three hundred or so of these tablets which have mathematical relevance, and discovered quite a wealth of achievement. Their published work since 1935 has shown a remarkably advanced mathematical understanding contemporary with the other ancient civilizations of China, India and Egypt, but much superior to any of these.

The tablets can be dated mostly to the time of the Sumerians (*c.*3000-2000 BC) and the Chaldeans (*c.*700-500 BC), with the major mathematical work belonging to the older period.

Numbers and Arithmetic

The first remarkable feature is the number system. The wedge-shaped characters use a vertical stroke $\vee$ for one and a horizontal stroke $<$ for ten, and small numbers are written additively, e.g. $<<< \vee\vee\vee\vee\vee$ for 35, as in the Egyptian system. As the numbers increase, a significant change occurs. In one tablet, for example, which gives a table of squares, the numbers are written in sequence starting 1, 4, 9, 16, 25, 36, 49, 1 4, 1 21, 1 40, 2 1, ... The change of notation used after the 49 indicates that the number 60 is being used as the base of a multiplicative system. A number like 2 1 then means $2 \times 60 + 1 = 121$.

When numbers higher than 60^2 are reached the place system is extended with a number like $70^2 = 4{,}900$ being represented by 1 21 40 or $1 \times 60^2 + 21 \times 60 + 40$. This is a remarkable discovery of the value and convenience of the place system for representing large numbers. The Babylonians used the same multiplicative principle on which our number system is based, in contrast to the additive method of the Egyptians, Greeks and Romans.

The choice of 60 as base is unique and in writing fractions has the advantage over the use of 10 in that 60 has more factors. The Babylonians exploited this advantage by developing the fractions in a way analogous to our own method of decimals. Fractions like $\frac{1}{2}, \frac{1}{3}, \frac{1}{4}, \frac{1}{5}, \frac{1}{6}, \ldots$ could be expressed in terms of sixtieths as $30'$, $20'$, $15'$, $12'$, $10'$, ... and the system would be extended to expressing fractions in terms of

$$\frac{1}{60} \quad \frac{1}{60^2} \quad \frac{1}{60^3} \quad \frac{1}{60^4}$$

etc. For convenience, we will write $\frac{n}{60}$ as n' or $\frac{n}{60^2}$ as n'', etc. Thus a fraction written $7'30''$ would represent $\frac{7}{60} + \frac{30}{60^2} = \frac{450}{60^2} = \frac{1}{8}$. The extent to which the place system for fractions was used can be seen from a table of reciprocals which begins:

1: 1	1
1: 1 0 16 53 53 20	59 43 10 50 52 48
1: 1 0 40 53 20	59 19 34 13 7 30

and continues extensively until

1: 3	20

is reached.

It should be mentioned here that the earlier tablets had no symbol for zero or at any rate the absence of a number in a place position. The existence of a zero can only be discovered from the context of a problem as in the above table. These reciprocal tables were used for purposes of division. Since $a \div b$ can also be expressed as $a \times \frac{1}{b}$ and multiplication is easier than division, division can be avoided if tables of reciprocals are available. Thus $3 \div 8$ becomes $3 \times 7'30'' = 22'30''$.

In addition to reciprocals, extensive tables of squares, cubes, sums of squares and cubes, etc., were calculated. The table of squares could be used inversely to find square roots but there is evidence that a much more sophisticated method of finding square roots was known. One of the tablets gives an approximation for $\sqrt{2}$ as $1\ 24'\ 51''\ 10'''$ equivalent to 1.4142 to four decimal places. This is obtained by the device of starting with a trial solution x. If x is too large, then $\frac{2}{x}$ will be too small since $x\frac{2}{x} = 2$, and conversely. In either case the average of x and $\frac{2}{x}$, that is $\frac{1}{2}(x + \frac{2}{x})$ will be the solution. Starting from 1, this process gives the successive iterations $1, 1\ 30', 1\ 25', 1\ 24'51''11'''$, $1\ 24'51''10'''$.

Algebra

There is considerable evidence that the Babylonians moved beyond proficiency in arithmetic into the field of basic algebra. Here is a literal translation from one of their tablets which indicates also the style of presentation. A problem is stated, the calculation performed and the solution given in an effective way, but without any notation, lay-out or appeal to general principles. *"I have added the surface and side of my square: 45'. You will put 1, unity. Break one in two: 30′. Multiply 30′ and 30′: 15′. Add 15′ to 45′: 1. That is the square of 1. Subtract 30′ which you multiplied from 1: 30′ the side of the square."*

It would appear from the solution that the problem is about finding the side of a square, so that the first sentence could be translated as a statement of the quadratic equation: $x^2+x = \frac{3}{4}$. The solution then follows in effect: take half of 1, square it and add to both sides, i.e. $x^2+x+\frac{1}{4} = \frac{3}{4}+\frac{1}{4} = 1$. Take the square root of each side: $x+\frac{1}{2} = 1$

and $x = \frac{1}{2}$ which is correct.

The method of completing the square to solve a quadratic has been implied and the problem could be re-stated theoretically as:

$$\text{if } x^2+ax = b, \text{ then } x = \sqrt{[(\tfrac{a}{2})^2+b]}-\tfrac{a}{2}.$$

Other problems equivalent to quadratic equations are solved by this algorithm, for example: "Surface less side of square 14.30. Take 1, divided into two parts 30′×30′ = 15′. You add to 14.30 and 14.30.15′ has root 29.30′. You add to 29.30′ the 30′ which you have multiplied by itself and 30′ is the side of the square."

After quadratics, even cubic equations are attempted! One problem when expressed in modern symbolism becomes: $x^2(12x+1) = 1\ 45'$. The solution is, in effect: multiply both sides by 12^2

$$(12x)^2\,(12x+1) = 4\ 12$$
$$12x = 6$$
$$x = 30'$$

This is a truly remarkable solution as the standard method for solving a cubic equation invented by Tartaglia in about 1540 AD is much more complicated! However, the Babylonian solution is not a genuinely algebraic one, only an ingenious use of tables. The device of multiplying 12^2 served to change the expression into a standard form, with the left-hand side becoming the sum of the cube and square of a certain number. Then to solve the problem all that is required is a table of values of n^3+n^2 for each number n. From such a table it can be seen that when $n^3+n^2 = 252$, $n = 6$.

A more legitimate algebraic procedure is provided in an example of the solution of a problem involving simultaneous quadratic equations.

"I have multiplied length and width thus obtaining the area. Then I added to the area the excess of the length over the width 3 3. Moreover I have added length and width 27. Required length and width."

The problem is equivalent to:

$$xy+x-y = 3\ 3$$
$$x+y = 27$$

The procedure to the solution is given as follows:

$$27+3\ 3 = 3\ 30$$
$$2+27 = 29$$

take one half of 29 (14 30′) 14 30′×14 30′ = 3 30 15′

3 30 15′−3 30 = 15′

square root of 15′ is 30′

14 30′+30′ = 15 length

14 30′−30′ = 14 "width"

subtract 2 which has been added to 27, from 14.

12, the actual width.

The first two lines of the solution are in effect,

$$xy+2x = 3\ 30$$
$$x+y+2 = 27$$

Why is the 2 added to the second equation? The answer to this question is found at the end of the problem where 2 is subtracted from a fictitious width to obtain the actual width. In effect a substitution has been made, $y' = y+2$ so that having found y', 2 is subtracted to obtain y. This substitution changes the two original equations to:

$$xy' = 3\ 30$$
$$x+y' = 27$$

which becomes a standard form, solved by a routine procedure. The routine appears to be: given $xy = P$, $x+y = a$, formulate:

$$\sqrt{[(\tfrac{a}{2})^2-P.]} \quad \text{Then } x = \sqrt{(\tfrac{a}{2})^2-P}+\tfrac{a}{2}$$
$$y = \sqrt{[(\tfrac{a}{2})^2-P]}-\tfrac{a}{2}$$

Similar techniques are used in more difficult problems involving more than one variable. Examples of such problems are:

$$x^2+y^2 = 25\ 25$$
$$y = 40'x+5$$
$$x^2+y^2+z^2 = 23\ 20$$
$$x-y = 10$$
$$y-z = 10$$

$$20'(x+y)-1'(x-y)^2 = 15$$
$$xy = 10.$$

Other algebraic devices known to the Babylonians were the summation of arithmetic and geometric progressions, and even the sum of the squares of successive integers. Unlike the Egyptians there appears no immediate application of the type of algebra developed here, with the possible exception of astronomical calculations. The Babylonian astronomers made remarkably accurate recordings of the positions of bodies in the solar system, and used first and second differences in order to interpolate and extrapolate the future positions of planets.

Geometry

In some respects Babylonian geometry did not advance beyond that of the Egyptians. They used geometry for practical rather than theoretical reasons and knew empirically the correct rules for the area of a triangle and trapezium as well as the volume of a cylinder. They erroneously took π as 3 but later amended it to $3\frac{1}{8}$.

There are, however, two problems which indicate a more advanced knowledge. In one of these: "a beam of length 30 stands against a wall. The upper end has slipped down a distance 6. How far did the lower end move?" The solution is given correctly as 18, indicating a knowledge of Pythagoras' theorem.

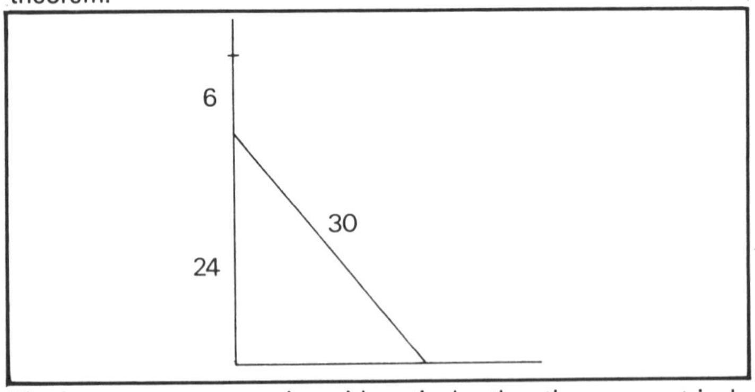

The interest appears to be arithmetical rather than geometrical, for another tablet gives a set of Pythagorean triads. Some are quite simple like 3,4,5 but others are more elaborate such as: 3 45, 3 31 39′, 5 9 1′.

It has been suggested that these triads were derived from knowledge of a general rule that if p,q are any numbers,

$$x = p^2-q^2,\ y = 2pq,\ z = p^2+q^2, \text{ then } z^2 = x^2+y^2.$$

The other geometrical result, one of the first of the mathematical texts to be deciphered, concerns finding the length of a chord of a circle given its "sagitta".

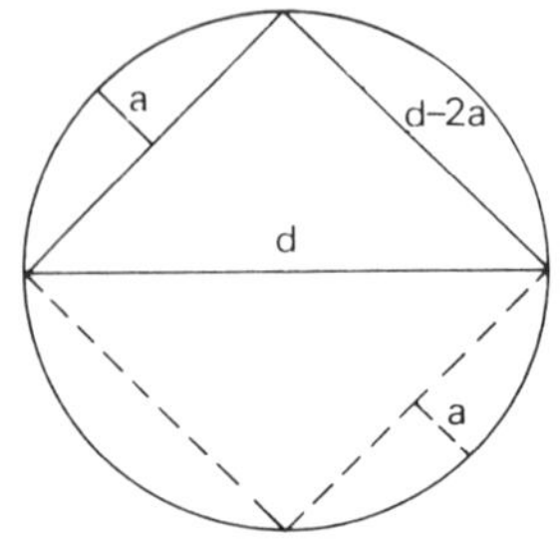

The result given is $\sqrt{d^2-(d-2a)^2}$ which shows that not only is Pythagoras' theorem known, but also the fact that the angle subtended by a diameter is a right angle.

It can be seen then, that the Babylonians had quite an advanced knowledge of Mathematics as early as 2000 BC. The continued examination of their records might reveal even more than has been expressed here. It has to be said, however, that what the Babylonians discovered largely perished with the fall of their civilization. Their algebraic methods were forgotten and had to be re-discovered centuries later. All that has remained is part of their sexagesimal number system which has proved to be very useful for time and angle measurement.

HISTORICAL TOPICS

ARITHMETIC SYMBOLS AND ALGEBRAIC NOTATION

by D. R. Green, CAMET, Loughborough University

This is the first of a series of articles which will trace the development of mathematics by examining the history of certain topics.

Although the four rules of arithmetic were well known in ancient times, the particular symbols which we use today are of much more recent origin. In the 15th century the Latin words "plus" and "minus" were used, and these were often abbreviated to "p̃" or "p̄" and "m̃" or "m̄". The bar "¯" signified that something had been omitted, i.e. was missing. Some writers (e.g. T. Dantzig, *Number the Language of Science*) claim that our minus sign "−" originated from replacing "m̄" by "−", dropping the letter "m". However, this neat idea is probably incorrect. In the middle ages the dash symbol "−" was often used to represent the letter "m" and it is feasible that this is how our minus sign originated. Merchants of the 15th and 16th centuries frequently indicated deficiencies in quantities of their goods by means of a dash. For example, "13 lb −4 oz" would mean 13 lb less 4 oz. In cases where there was excess, no sign was included, thus "13 lb 4 oz" would mean 13 lb and 4 oz. So these two ideas "−" for "m" and "−" for deficiency brought about the sign "−" for the operation of subtraction.

The plus sign has a rather different origin, coming from an abbreviation of the Latin word "et" meaning "and". Cajori relates that 102 different abbreviations for "et" have been recorded so it is not surprising that one of these, first appearing in 1417, is similar to "+"!

The German mathematician Johann Widman (1460?-1500?) was the first to put "+" and "−" into print in his arithmetic book published 1489, although there they are used to represent excesses and deficiencies rather than the operations of addition and subtraction (Figure 1).

The "+" and "−" signs had occurred earlier than this in manuscripts on algebra (1481 for example). Indeed Widman had seen the 1481 manuscript, and students attending his lectures at Leipzig University in the 1480s used the new notation.

Various European mathematicians used the German signs "+" and "−" thereafter, firstly in algebra and then in arithmetic, and it was the important German algebraist, Michael Stifel (ca 1487-1567) in his arithmetic books (1544, 1545) who did most to popularise the two symbols. The Italian algebraists continued in their Latin tradition by using "p̄" and "m̄" but the English quickly adopted the new Germanic signs, and indeed Robert Recorde (1510-1558) used them in his famous arithmetic and algebra books, the *Grounde of Artes* (1542?) and the *Whetstone of Witte* (1557). In being written in English rather than in Latin, they did much to bring mathematics within reach of many whose education did not allow them to master the classical writings. Recorde himself only invented one symbol which has proved successful and that was the sign for equality. The sign was introduced in the *Whetstone of Witte* (1557) in a form longer than used today and was an "arbitrary" invention, symbolising a pair of parallel lines. Recorde explains his reasons for the choice — see Figure 2. (Note: "Gemowe" means "twin".)

Figure 1

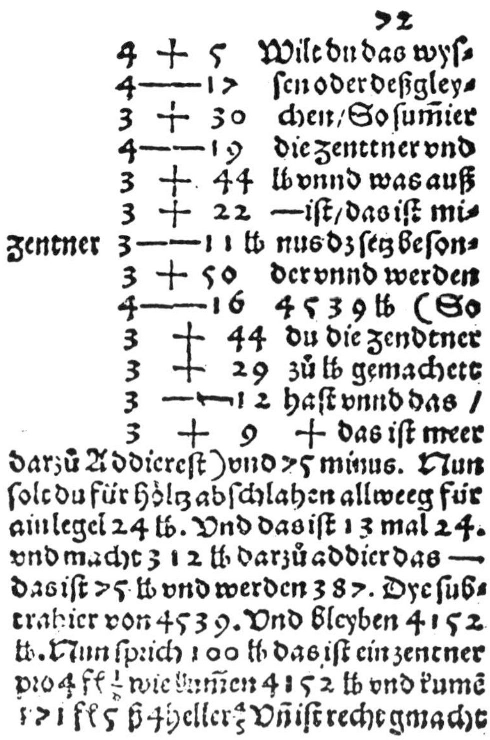
72
4 + 5 Wiltu dis das wyſ-
4 — 17 ſen oder deßgley-
3 + 30 chen / So ſumier
4 — 19 die zentner vnd
3 + 44 lb vnnd was auß
3 + 22 — iſt / das iſt mi-
zentner 3 — 11 lb nus d3 ſetz beſon-
3 + 50 der vnnd werden
4 — 16 4539 lb (So
3 + 44 du die zendtner
3 + 29 zů lb gemachet
3 — 12 haſt vnnd das /
3 + 9 + das iſt meer
darzů Addiereſt) vnd 75 minus. Nun
ſolt du für holtz abſchlahen allweeg für
ain legel 24 lb. Vnd das iſt 13 mal 24.
vnd macht 312 lb darzů addier das —
das iſt 75 lb vnd werden 387. Dye ſub-
trahier von 4539. Vnd bleyben 4152
lb. Nun ſprich 100 lb das iſt ein zentner
pro 4 ff $\frac{1}{5}$ wie kummen 4152 lb vnd kume
171 ff 5 ß 4 heller $\frac{2}{5}$ Vnd iſt recht gemacht

Figure 2

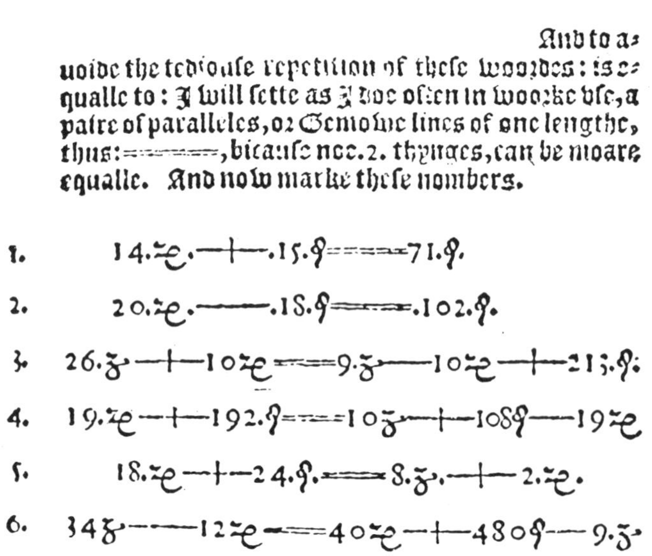
And to a-
uoide the tediouſe repetition of theſe woordes : is e-
qualle to : I will ſette as I doe often in woorke uſe, a
paire of paralleles, or Gemowe lines of one lengthe,
thus: ═══, bicauſe noe. 2. thynges, can be moare
equalle. And now marke theſe nombers.

1. 14.ϛ.—+—.15.ϕ.═══71.ϕ.
2. 20.ϛ.———.18.ϕ.═══.102.ϕ.
3. 26.ʒ.—+—10ϛ═══9.ʒ.——10ϛ—+—213.ϕ.
4. 19.ϛ—+—192.ϕ.═══10ʒ—+—108ϕ——19ϛ
5. 18.ϛ—+—24.ϕ.═══8.ʒ.—+—2.ϛ.
6. 34ʒ———12ϛ═══40ϛ—+—480ϕ——9.ʒ.

In modern type Recorde's explanation appears as follows:
And to avoide the tediouse repetition of these woordes :is equalle to: I will sette as I doe often in woorke use, a paire of paralleles, or Gemowe lines of one lengthe, thus:═══, because noe. 2. thynges, can be moare equalle. And now marke these numbers.
1. 14.cwt. —+— . 15.Qtr ═══ 71 Qtr.

....

However, various other signs for equality, especially the older sign "∞", itself possibly an abbreviation for the Latin word "aequales", continued in vogue for another two centuries and the (shrinking!) sign for equality "=" only gradually took over. Cajori remarks that it was a close thing as to whether "=" or "∞" would finally triumph and Leibniz's preference for "=" was probably the decisive factor.

The exact derivation of the symbol "×" for multiplication is not so easily discerned. Various large crosses were used in setting out certain operations with fractions. For example, Cajori refers to Borgi dividing $\frac{3}{4}$ by $\frac{4}{5}$ setting it out thus:

4 ╲╱ 3 15
– ╳ – —
5 ╱╲ 4 16 (Borgi, 1488)

Pacioli, the Italian friar, set out the squaring of the two-digit number 37 thus:

3 7
| ╳ |
3 7
———
1 3 6 9 (Pacioli, 1494)

where the four upper lines indicated which digits must be multiplied together. Sometimes in similar multiplications the two vertical lines were omitted thus:

7 8
╳
5 6
———
4 3 6 8 (Unicorno, 1598)

somewhat suggestive of the multiplication sign "×".

The credit for introducing a more normal sized times sign, truly indicating the operation of multiplication, usually goes to the Englishman William Oughtred (1575-1660) but not for some three centuries after his introducing it in 1631 did its widespread use in elementary arithmetic occur. An objection raised to using "×", which remains valid today, was its possible confusion in algebraic work with the letter x, and so the dot symbol was favoured. The dot had been employed for some time but more as a separator than an explicit operation sign and often no actual symbol for multiplication was used, juxtaposition of the variables being sufficient. Smith states that the important English mathematician Thomas Harriot (1560-1621) used the dot for multiplication in cases such as "2.aaa" for "$2a^3$" but this is more probably an example of the dot merely signifying the end of the numeral 2. It is probably true to say that Leibniz invented the dot for multiplication in 1698 when he raised the objection to "×"mentioned above.

The division sign "÷" has a similar history to that of "−", being an old merchant's sign gradually adopted by algebraists. Smith records that Lombard merchants used "÷" to mean $\frac{1}{2}$, so $2\frac{1}{2}$ was written "2÷". Despite the widespread use of "÷" to represent subtraction, it was proposed as a symbol for division by the Swiss Johann Rahn (1622-1676). The first appearance in print of "÷" representing division occurs in Rahn's *Teutsche Algebra* published 1659 (Figure 3).

Figure 3

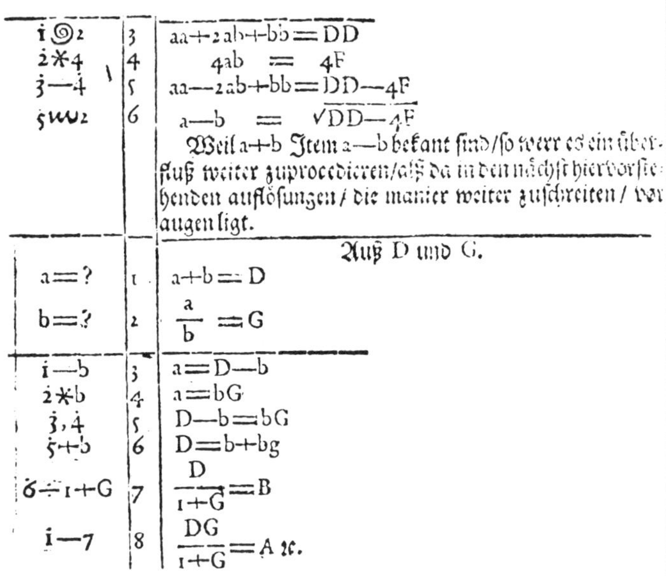

The *Teutsche Algebra* was translated into English by Brancker and published, with some additions by John Pell, in 1668 and so the symbol came to England where it was widely adopted. However, the continental mathematicians preferred the ":" sign.

Our long division symbol ")" was used by Stifel (1544) as in 8)24 representing 24÷8, and by Oughtred (1631) as in 8)24(3.

The inequality symbols are of varied heritage. The two basic symbols "<" and ">" were invented by Harriot and published posthumously in 1631, and were used by Oughtred in 1677 for example. However, the pair "≦" and "≧" are attributed to a little-known French mathematician P. Bouguer (1734). Smith reports that the striking through a relational symbol to denote its negation (e.g. ≮ meaning 'not less than') is a modern idea.

For many centuries the common root symbol used was "℞" standing for the Latin word "Radix", itself a translation of the Arabic word "jidr" meaning a plant root and used by Arab mathematicians for the square root. It was necessary for various roots, not only square roots, to be indicated and a variety of symbols were devised, usually consisting of "℞" followed by a small numeral or an appropriate word probably abbreviated. Thus the cube root sign might be:

"℞ 3ª"	Chuquet	1484	(French)
"℞3ª"	Pacioli	1494	(Italian)
"℞ cub."	Cardan	1545	(Italian)

The modern "√" sign, although introduced as early as 1525 in Germany by Christoff Rudolff (1499?-1545?) in his algebra book *Die Coss*, (Figure 4) only became standardised into its modern usage after many variations had been experimented with.

Rudolff's own notation was rather clumsy and confusing in that "√" was square root, "√√" was fourth root (really representing "√√") and "√√√" was cube root. Among the notations used for the cube root were:

√√√	Rudolff	1525	(German)
∛	Stifel	1545	(German)
·√√√·	Recorde	1557	(English)
∛	Girard	1629	(Lorraine)
√3.)	Harriot	1631	(English)
√ ③	Stevin	1634	(Belgian)
√C	Descartes	1637	(French)
$\sqrt{}^{3}$	Wallis	1655	(English)

It is seen that Girard was first to use our notation but not until the 18th century did it become widespread. It has been suggested (but never substantiated) that "√" is a corruption of the letter "r". Incidentally, the title of Rudolff's algebra book, namely *Die Coss*, literally meant "the thing" because the unknown quantity in algebra (which we frequently refer to as x) was known by the name "coss" in German, being a transliteration of the Italian word "cosa" itself the translation of the Latin word "res" meaning "thing" which was used for the unknown in Latin algebra works. This also explains why the subject algebra was known in the 16th and 17th centuries in England as the "Cossic Art".

The word "surd" — often used to describe roots not expressible as rational numbers — literally meant "deaf" or "mute", referring in a way to their inexpressibility. This idea originated with Arab mathematicians of the 9th century, who were so important in transmitting the culture of the Greeks to Western Europe.

From ancient times powers of unknowns have been needed in algebra problems — equivalent to our x^2, x^3, etc — and originally quite different names were used to designate each separate power. Later on repetition of the unknown, by now abbreviated to a letter, was employed, thus a^3 would be written "*aaa*", as used by Harriot in the early 17th century for example. Many notations were tried out in order to simplify the writing of powers and then in 1637 Descartes (1596-1661) invented the present exponential notation as a logical extension of the best efforts current in his day. However, it took another hundred years before his notation was widely adopted. In the meantime Newton (1642-1727), for example, would write:

$$zzxx = aapp - p^2x^2$$

One final symbol to be considered here is the "%" sign, which developed from the various abbreviations used in the 15th to 17th centuries for the Latin words "per cento" meaning "out of a hundred". Some of the stages in its development were:

per cēto

p cento

p c°

per $\frac{\circ}{\circ}$

and by the middle of the 17th century the "per" was dropped, leaving "$\frac{\circ}{\circ}$" which with modern printing has become "%".

Figure 4

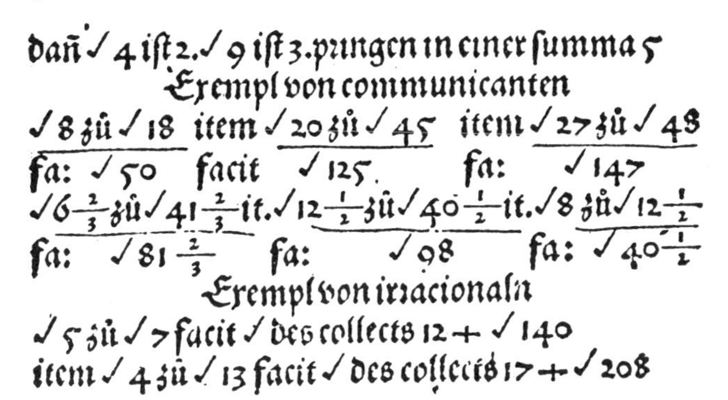

BIBLIOGRAPHY

History of Mathematics Vol. II, D. E. Smith (1925 Edition), Dover, 1953.

A History of Mathematical Notations, Vol. I, F. Cajori, Open Court, 1928. (The publishers' permission to reproduce figures is gratefully acknowledged.)

by D. R. Green, CAMET Loughborough University

PLANE TRIGONOMETRY

There is no absolutely sure beginning to what we call trigonometry and opinions vary from around 200 B.C. to A.D. 1600! However, it is generally acknowledged that the earliest known precursor of the subject is found in Egyptian civilisation. Appropriately, this relates to those great monuments of their culture — the Pyramids.

The Origins of Trigonometry

The Egyptians

Generally speaking the pyramids were built on a square base with each of the four triangular faces inclined at an angle of about 52° to the base (Fig. 1).

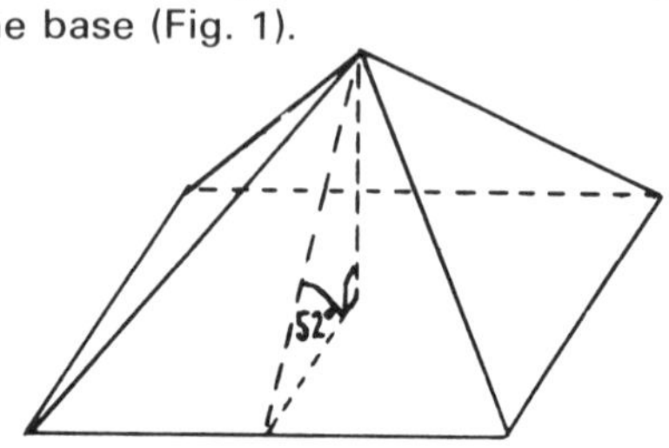

Fig. 1

The most important early Egyptian mathematical document is the Rhind Papyrus which is now housed in the British Museum. It is named after Henry Rhind who brought it to Europe. Also known as the Ahmes Papyrus, after the scribe who prepared it around 1650 B.C., it is a scroll about 5½m long by 30cm wide containing some 84 problems with their solutions. Problems 56 to 60 deal with mensuration of pyramids and four of these make use of the term *seqt* closely related to the cotangent. The *seqt* was the horizontal change for each unit increase in vertical height (Fig. 2).

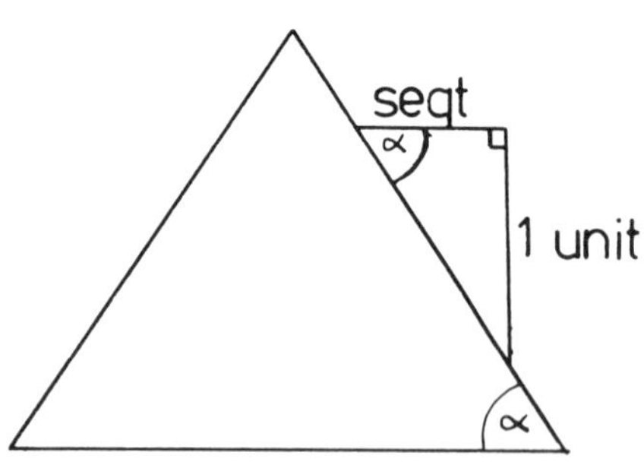

Fig. 2

To maintain a uniform slope — giving a flat face to the pyramid — it was necessary to ensure that the *seqt* was constant. A slight complication, as far as we are concerned, is that the *vertical* height was in *ells* but the *horizontal* distance was in *hands* (1 ell = 7 hands).

Problem 56 asks: "Find the seqt of a pyramid which has a square base of side 360 ells and a vertical height of 250 ells". The answer given is a seqt of $5^{1}/_{25}$ (hands per ell). The reader may like to check this.

The famous Cheops Pyramid, so Boyer (Ref. 1) tells us, has dimensions of 440 ells width, 280 ells height. What is the seqt? What angle does an edge between two adjacent sloping faces make with the base? (Ref. 2).

Basically, then, the *seqt* was cot α, where α is the angle of slope of each face of the pyramid, but because of the units used:

$$seqt = 7 \cot \alpha$$

It is easy to envisage the Egyptian surveyors checking the construction of the pyramids with simple apparatus using this principle.

The Greeks

Trigonometry as we know it really began with the Greeks in their astronomical investigations. (Not until the Arab Nasir Eddin (1201-1274) did trigonometry actually appear separated from astronomy). The simplest piece of astronomical apparatus was the vertical stick, or *gnomon*, used to produce a shadow on the horizontal ground. By careful measurement of the shadow length — hour by hour and day by day — the time of year and time of day could be determined (Fig. 3).

Fig. 3

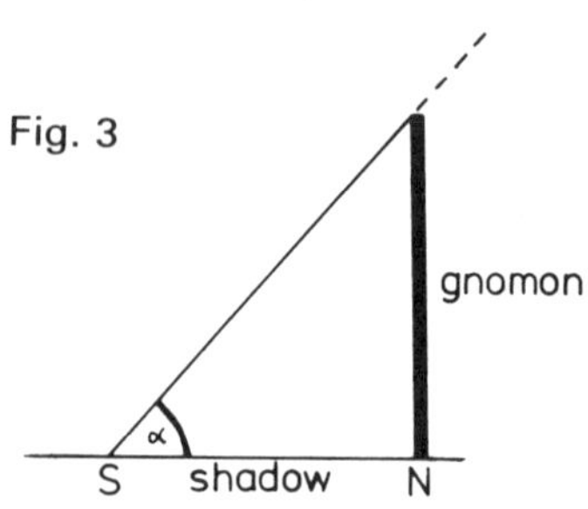

The varying shadow length, SN, is entirely dependent on the angle (α) made by the sun: SN = GN cot α. Egyptians and Babylonians (c. 1500 B.C.) and Chinese (c. 1100 B.C.) are known to have used such devices too but it was the Greeks who went further by studying the geometry of planetary motion, thereby bringing in trigonometry.

Basic to these calculations was the relationship between the *chord* and the *angle* in a circle (Fig. 4).

Fig. 4

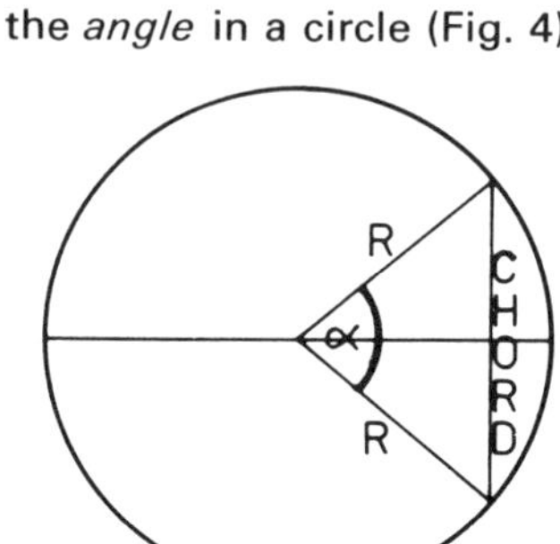

The Greeks dealt in chords rather than sines but there is a simple relationship:

$$\text{chord }(\alpha) = 2R \sin \left(\frac{\alpha}{2}\right)$$

Tables of chords, rather like our tables of sines, were produced by Greek astronomers (e.g. Menelaus c. 100) for some chosen radius R (not R = 1 as we choose for sine). Ratios of lengths were much used by the Greeks to express theorems. W. Popp (Ref. 3, p. 66) gives a good example: Archimedes (287 B.C.-212 B.C.) gave the result

$$AC:DC = AB:BC + AC:BC \qquad \text{(see Fig. 5)}$$

Fig. 5

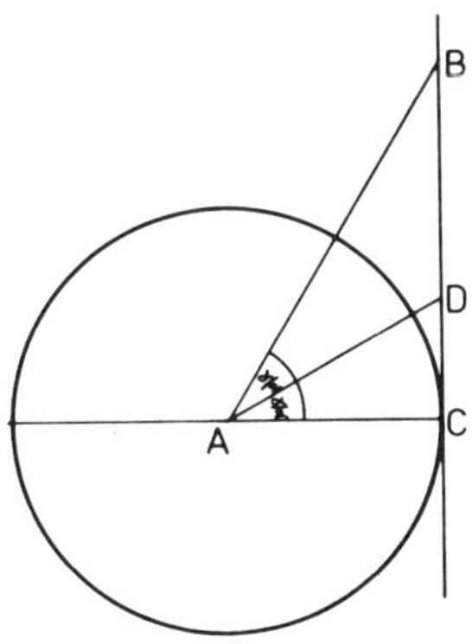

The reader may care to show that this is equivalent to

$$\cot\frac{\alpha}{2} = \frac{1}{\sin\alpha} + \cot\alpha$$

which is indeed a trigonometrical identity as the reader can check.

Perhaps the most famous astronomical treatise was "Almagest" of Ptolemy (c. 85-c.165). At the beginning of this great work are various trigonometrical formulae leading to the calculation of a table of chords for values of ½° to 180° in steps of ½°. The "Almagest" (meaning "the Greatest") as the Arab translators called it, was essentially a summary of the Theorems known to Hipparchus (c. 140 B.C.).

Ptolemy's Theorem, as it is known, states that for a cyclic quadrilateral ABCD

$$BC \times AD = AC \times BD - AB \times CD$$

This was well known to the Greeks, and Ptolemy applied it to the special case where AD is the diameter of the circle (Fig. 6).

Fig. 6

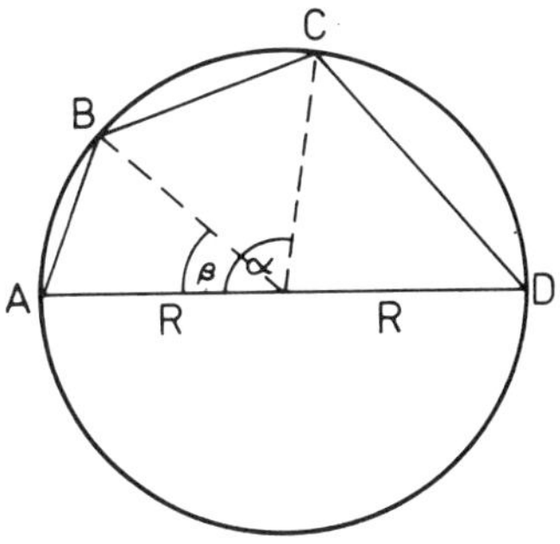

The result is then:

$$\text{chord }(\alpha-\beta) \times 2R = \text{chord }(\alpha) \times \text{chord }(180°-\beta) - \text{chord }(\beta) \times \text{chord }(180°-\alpha) \qquad \text{I}$$

Now in our modern notation, using sin and cos we know that chord (x) = $2R\,.\sin\left(\frac{x}{2}\right)$

so, taking $A = \frac{\alpha}{2}$, $B = \frac{\beta}{2}$ the identity is

$$2R\sin(A-B) \times 2R = 2R\sin A \times 2R\sin(180°-B) - 2R\sin B \times 2R\sin(180°-A)$$

i.e. $\sin(A-B) = \sin A\cos B - \sin B\cos A$, a familiar result.

Ptolemy similarly derived the chord formula equivalent to the identity $\sin(A+B) = \sin A\cos B + \sin B\cos A$, namely:

$$\text{chord }(\alpha+\beta) \times 2R = \text{chord }(\alpha) \times \text{chord }(180°-\beta) + \text{chord }(\beta) \times \text{chord }(180°-\alpha) \qquad \text{II}$$

These results were used by Ptolemy to compute tables. The circle was taken to have a radius R = 60 *parts* (called *moiras* in Greek and *degrees* in Latin). Each *moira* was equal to 60 *minutes*, each *minute* equal to 60 *seconds* . . . The use of a sexagesimal base was the influence of the Babylonian system. With the commonly taken value of $\pi \simeq 3$ this naturally gave 360 *parts* for the circle circumference, tallying with our normal 360° representation of a full turn. It must be emphasised that these *degrees* or *moiras*, as the *parts* were called, are units of length and not of angle. To make this clearer the symbol p will be used instead of our more normal degree symbol, but the subdivisions into minutes and seconds will be given the familiar ′ and ″ symbols. From simple geometrical considerations of regular inscribed polygons (Fig. 7)

Fig. 7

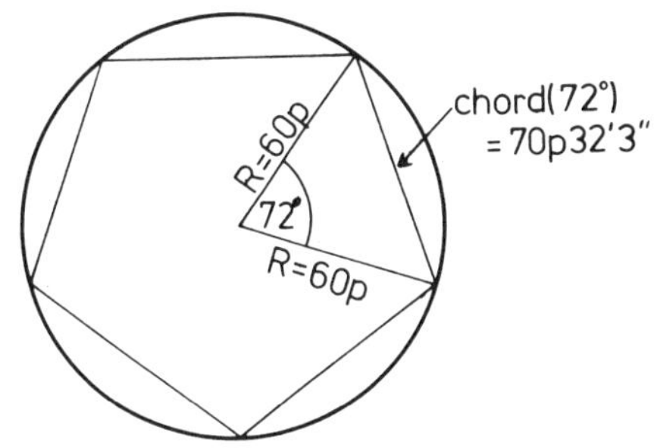

Ptolemy showed that

10 sides: chord(36°) ≃ 37 p 4′ 55″ i.e. $37 + \frac{4}{60} + \frac{55}{60^2}$ parts

6 sides: chord(60°) = 60 p

5 sides: chord(72°) ≃ 70 p 32′ 3″

4 sides: chord(90°) ≃ 84 p 51′ 10″

3 sides: chord(120°) ≃ 103 p 55′ 23″

It is interesting to check the accuracy of Ptolemy's basic results:

For example

chord(36°) =	2×60×sin(18°)	= 37.08204 to 5 d.p.
Ptolemy's result:	37 p 4′ 55″	= 37.08194 to 5 d.p.

so the result is clearly correct to 5 sig. fig.

Further values, for chord (108°) and chord (144°), were easily derived directly from the results for chord (72°) and chord (36°) respectively, using Pythagoras' Theorem. (Fig. 8).

e.g. chord(180° − 72°) = chord(108°) = $\sqrt{120^2 - \text{chord}(72°)^2}$

≃ 97 p 4′ 55″

Fig. 8

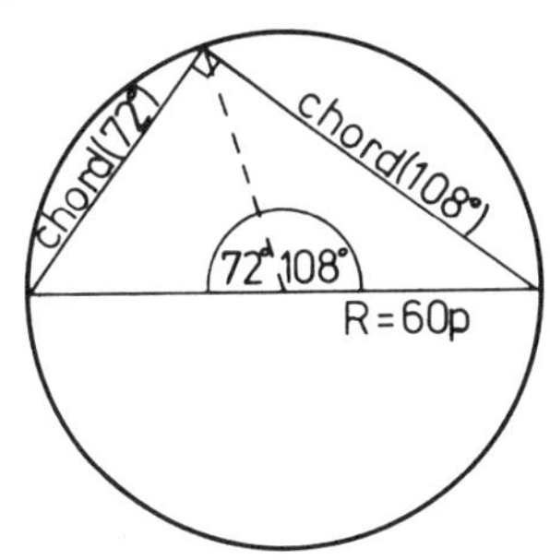

The result for chord(144°) is left for the reader to find (Ref. 4). From the results for chord(60°) and chord(72°) Ptolemy applied his formula I to derive chord(12°)

= chord(72° − 60°)

$$= \frac{1}{120}\left[\text{chord}(72°) \times \text{chord}(60°) - \text{chord}(60°) \times \text{chord}(36°)\right]$$

≃ 12 p 32′ 36″

Now Ptolemy really wanted to find chord(½°) and thus chord(179½°) from which to build up the complete table in stages of ½°, using

$$2 \times 60 \text{ chord}(\alpha + ½°) = \text{chord}(\alpha)\ \text{chord}(179½°) + \text{chord}(½°)\ \text{chord}(180° - \alpha)$$

To do this he used the equivalent to the half-angle formula $\sin\frac{\alpha}{2} = \sqrt{(1 - \cos\alpha)/2}$, which was:

$$\text{chord}\left(\frac{\alpha}{2}\right) = \sqrt{(120^2 - 120\ \text{chord}(180 - \alpha)/2)} \qquad \text{III}$$

(see Ref. 1, p. 186 for details)

Thus from chord(12°) Ptolemy could get values for 6°, 3°, 1½°, ¾°. By estimating ("linear interpolation") from the values for 1½° (1 p 34′ 15″) and ¾° (0 p 47′ 18″) he deduced that chord(1°) ≃ 1 p 2′ 50″. The interested reader can derive these results and check their accuracy.

Having found chord(1°) ≃ 1 p 2′ 50″, knowing various chord values and having the chord formulae II and III, it was straightforward (if somewhat tedious!) to compile a table of chord values for ½° up to 180° (See Ref. 5, pp. 127-137 for details).

The Hindus

Hindu astronomers, somewhat later, developed their mathematics along rather different lines. They took as the basis the *semi-chord* and the *half-angle.* Aryabhata (c. 500) drew up a semi-chord table at intervals of 3° 45′ which, apart from not taking R = 1, is effectively a table of sines.

The Arabs

The Arab mathematician-astronomers were conversant with both Greek and Hindu traditions and, recognising that the semi-chord approach of the Hindus was more useful, went on to develop the subject further. For example, a cotangent table giving the length of shadow of a stick produced by the sun at angle α for 1° intervals was drawn up by Habash al Hasib around 700-850. Another Arab, Abû'l-Wefâ (c. 980), made Ptolemy's Almagest widely known and gave great attention to accurately preparing tables and systematising known trigonometrical theorems. For example, his value for ½ chord(1°) (= 60 sin (½°)) was 0 p 31′ 24″ 55‴ 54⁗ 55′′′′′, as accurate a value as any pocket calculator is likely to give!

The Europeans

The Arabic work on trigonometry became known in Europe via Spain in the 13th and 14th Centuries. Then in the 15th Century G. Peurbach (1423-1461) and his pupil Regiomontanus (so called because he came from Königsberg) brought the subject to the fore with their preparation of new extensive trigonometrical tables to great accuracy, and with Regiomontanus' treatise "De triangulis omnimodis" (completed about 1464) which summarised the trigonometrical knowledge of the time (see Ref. 5, pp. 196-200 and Ref. 6, pp. 144-145, for details). Regiomontanus (real name Johannes Müller, 1436-1476) used R = 6 000 000 and later R = 10 000 000 in order to avoid working with decimal fractions and yet get very accurate values. He was well acquainted with both Arabic and Greek astronomical works which made use of trigonometrical calculations and to him is given much credit in creating trigonometry as a subject in its own right, following in the tradition of the Arab Nasir Eddin.

A subsequent great feat of table preparation — and perhaps one never since surpassed — was that undertaken by Rhaeticus (1514-1576) who, in 1551, published tables for all six trigonometric functions, at 10′ intervals, to 7 sig. fig., and, not content with that, Rhaeticus began a new set of tables this time at 10″ intervals, to 10 sig. fig. This was completed and published long after Rhaeticus' death by his pupil V. Otho, in 1596.

The developments which followed were mainly in the direction of linking trigonometry with algebra and then with complex number theory. An early pioneer in this, and a great preparer of tables too, was the French mathematician François Viète (1540-1603), often known as Vieta. The use of various values for the radius R was slow to disappear despite the fact that decimal fractions had been invented and were known. Not until 1748 when Euler gave his support to fixing R = 1 did the idea of the trigonometric functions being *ratios* rather than *lengths* gain wide acceptance. Smith (Ref. 7) records that the first to define the trigonometric functions as pure numbers was the German mathematician A. G. Kästner in 1759.

John Wallis (1616-1703) and Isaac Newton (1642-1727) were two great English mathematicians who contributed to analytic trigonometry, through the development of series expansions, and several Europeans investigated the relationship between complex numbers and trigonometrical functions: Jean Bernoulli in 1702 discovered that

$$\text{Arc}\sin x = -i \ln[(1 - x^2) + ix] \quad (x^2 \leq 1)$$

and De Moivre (1707), Cotes (1722) and finally Euler (1748) showed in varying forms that

$$e^{ix} = \cos x + i \sin x$$

thereby firmly linking the trigonometric and exponential functions together, through complex numbers.

The Origins of Trigonometrical Terms

Chord

As we have seen, the *chord* was the fundamental entity in the Greeks' trigonometrical work. Our word chord derives from the Latin "chorda" meaning a bowstring (quite suggestive of the shape produced by the chord and arc (Fig. 4)) which in turn comes from the Greek "chorde" which meant the intestine of an animal and hence a string for a bow.

Sine

Our preferred fundamental trigonometric entity is the sine which derives from the Hindu mathematicians as does the term itself, indirectly. The Hindu term "ardhā-jyā" meaning "half chord", which was what the sine was basically, was used by Aryabhata (c. 500). This became abbreviated to "jyā" and when Arabic translators came across this term it was transliterated to the invented word "jiba" which later was changed to "jaib" which meant bosom or bay. When European scholars in the 12th Century, and later, came to translate from the Arabic works they naturally used "sinus" which is the Latin for bosom or bay.

Versed Sine

The versed sine was much used by Hindus and then Arabs and then Europeans but is not now given any special name. It corresponds to the length of radius bounded by the chord and the arc (Fig. 9), and so is given by

$$\text{versin}(\alpha) = R\left(1 - \cos\frac{\alpha}{2}\right).$$

Fig. 9

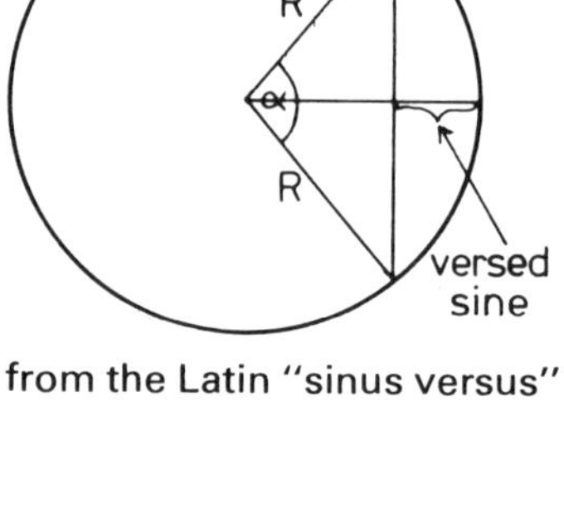

The term comes from the Latin "sinus versus" — the "turned sine".

Cosine

The Greeks had no need of the cosine and even though the

Hindus and Arabs used the idea they generally were content to express $\cos\theta$, in effect, as $\sin(90° - \theta)$. Regiomontanus (c.1463) used "sinus rectus complementi" and this eventually became abbreviated to the accepted forms: "co.sinus" by E. Gunter (1620) and then to "cosinus" by J. Newton (1658) and then "Cos." by J. Moore (1674).

Tangent and Cotangent

The gnomon and its shadow were the origin of these two trigonometrical ideas and were termed the "turned shadow" and "shadow" by the Arabic writers, and these terms became "umbra recta" and "umbra versa" in Latin and were widely used from the 13th Century onwards. The term "tangent" comes from the Latin verb "tangere" meaning to touch (an appropriate description of the relationship between the tangent and the circle — Fig. 10) and was first used in the form "tangens" by T. Fincke (1583) and made popular by Pitiscus (1595), famed for his trigonometrical tables with 15 place accuracy. Gunter (1620) gave us the Latin term "cotangens", and the abbreviation "tan", as in $\overset{\text{tan}}{A}$ for tan(A), is due to Girard (1626) while "Cot." was the suggestion of J. Moore (1674).

Secant and Cosecant

The concepts of secant and cosecant arose much later than those of the other trigonometrical functions, and Smith (Ref. 7) records their first appearance (without special name) in the works of the Arab Abû'l-Wefâ (c. 980) although even there little use was made of them. Only in 15th Century navigators' tables do they appear seriously. The first *printed* tables of these functions were those of Rhaeticus (1551). The Latin word "secans", coming from the verb "secere" meaning to cut (again appropriate, see Fig. 10), was suggested by Fincke (1583) and the abbreviation "sec", as in $\overset{\text{sec}}{A}$ for sec(A) is due to Girard (1626).

Fig. 10

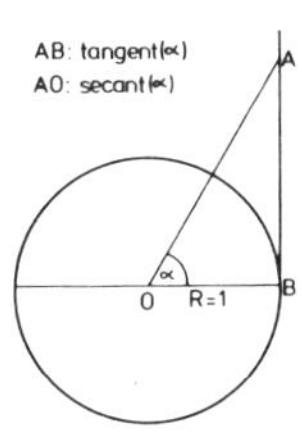

Trigonometry

This word comes from the Greek words tri (three), gonon (angles), metron (measure) and so signifies triangle measurement. It first appeared in the title of a book by B. Pitiscus, "Trigonometria", published in 1595.

Degree, minute, second

Our familiar angle units derive from Latin. The original Greek word used in trigonometry (but really indicating a length of course) was moira, or part, which was divided into subparts (sixtieths) and so on . . . The Latin translation of moira was degradus (step) and the subparts were: "pars minuta prima" (first small part) leading to our word minute and "pars minuta secunda" (second small part) leading to our word second. As has been indicated, 360° for a full turn was taken based on the Babylonian number system, with a base of 60. Precise details of why 360 was used is not known. Certainly a base of 60 makes many division processes very simple and fraction work easy.

Radian

In 19th Century England a physicist J. T. Thomson and a mathematician T. Muir independently came to the conclusion that a new angle measure was needed to simplify handling trigonometrical functions, especially in the context of calculus. They met and in consultation with A. Ellis agreed upon the name radian — 2π radians corresponding to 360° (so 1 radian $\simeq$ 57°17′45″). The first time the word radian appeared in print was in an examination paper set in 1873 by Thomson, and the first book to use the term was published by Ellis in 1874.

The Origins of Some Theorems of Plane Trigonometry

	Theorem	Known to:	Clearly stated by:
1	$\sin(\alpha \pm \beta) = \sin\alpha \cos\beta \pm \cos\alpha \sin\beta$	Ptolemy (c.150)	Hipparchus (c.140 B.C.)
2	$\sin 2\alpha = 2\sin\alpha\cos\alpha$	Hipparchus (c.140 B.C.)	Abû'l Wefâ (c.980)
3	$\sin 3\alpha = 3\sin\alpha - 4\sin^3\alpha$	—	Vieta (1591)
4	$\sin^2\alpha + \cos^2\alpha = 1$	Hipparchus (c.140 B.C.)	Varahamiriha (c.505)
5	$\sin\left(\frac{\alpha}{2}\right) = \sqrt{\frac{1-\cos\alpha}{2}}$	Ptolemy (c.150) or Hipparchus (c.140 B.C.)	Varahamiriha (c.505)
6	$a^2 + b^2 - 2ab\cos C = c^2$	Euclid (c.300 B.C.)	Vieta (1593 and others
7	$\frac{a}{\sin A} = \frac{b}{\sin B} = \frac{c}{\sin C}$	Ptolemy (c.150)	Nasir Eddin (c.1250)
8	Area of triangle = $\frac{1}{2}ab\sin C$	Regiomontanus (c.1464)	Snell (1627)
9	$\tan 2\alpha = \frac{2\tan\alpha}{1-\tan^2\alpha}$	—	Euler (1748)
10	$\cos 2\alpha = \frac{1-\tan^2\alpha}{1+\tan^2\alpha}$	—	Lambert (1765)
11	$\frac{a+b}{a-b} = \frac{\tan\frac{1}{2}(A+B)}{\tan\frac{1}{2}(A-B)}$	Fincke (1583)	Vieta (c.1590)

References

1 Boyer, C. B. *A History of Mathematics.* Wiley, 1968.
2 5½ hands per ell; 42°.
3 Popp, W. *History of Mathematics. Topics for Schools.* Transworld, 1975.
4 144 p 7′ 37″.
5 Dedron, P. & Itard, J. *Mathematics and Mathematicians, 2.* Transworld, 1974.
6 Dedron, P. & Itard, J. *Mathematics and Mathematicians, 1.* Transworld, 1974.
7 Smith, D. E. *History of Mathematics, Vol 2.,* Dover, 1958.

The history of vectors and matrices

by A. P. Knott, University of Technology, Loughborough

The concept of a vector is an old one; far older than the name. Artistotle knew that forces can be respresented by directed line segments and that addition of forces follows what is generally called the parallelogram law. Simon Stevin, who did so much to advance the course of decimals, used the parallelogram law frequently in solving problems of statics and the law was stated explicitly by Galileo.

With hindsight, we can recognise that the modern development of vectors starts with the geometrical treatment of complex numbers proposed by Argand and Wessel. By about 1830 complex numbers were frequently used to represent vector quantities and many mathematicians were undertaking the search for the three dimensional analogue of complex numbers. Wessel, Gauss, Servois and Möbius were some of the mathematicians who attempted this problem and none was successful. The credit for the development of such an analogue must go to the Irish mathematician, W. R. Hamilton. Hamilton was born in 1805 and for much of his youth his main interests appeared to be in languages — by the age of 14 he could speak nine languages. By the age of 17 however he had developed an interest in mathematics and had presented his first paper to the Royal Irish Academy. His mathematical interests were mainly in those areas of mathematics applicable to physics.

In 1833 he presented in a long paper before the Royal Irish Academy a formal treatment of complex numbers. Here, he defined a complex number as an ordered pair of real numbers subject to certain abstract rules of combination. Thus he defined multiplication as

$$(a, b) \cdot (c, d) = (ac - bd, ad + cd)$$

He naturally tried to extend this idea to three-dimensions. He thought that a three-dimensional analogue of a complex number would be a triple $a + b\underline{i} + c\underline{j}$ where $\underline{i}$ and $\underline{j}$ are both the square roots of -1 but distinct from one another. The operation of addition of triples proved no problems but for 10 years he was unable to multiply such objects. On and off the triples took up much of his life. In a letter to one of his sons he wrote[1]

"Every morning . . . , on my coming down to breakfast, your brother William Edwin and yourself used to ask me, 'Well, Papa, can you multiply triplets?' Whereto I was always obliged to reply with a sad shake of the head, 'No, I can only add and subtract them'."

When the solution came it came in a most surprising way. In a well documented flash of inspiration he saw that he needed to both abandon the commutative law of multiplication and use quadruples rather than triples. He wrote in 1865 to his son:[2]

"On the 16th day of October, which happened to be a Monday, and Council day of the Royal Irish Academy, I was walking in to attend and preside, and your mother was walking with me along the Royal Canal, to which she had perhaps driven; and although she talked with me now and then, yet an undercurrent was going on in my mind, which gave at last a result, whereof it is not too much to say that I felt at once the importance. An electric circuit seemed to close; and a spark flashed forth, the herald (as I foresaw immediately) of many long years to come of definitely directed thought and work, by myself if spared, and at all events on the part of others, if I should even be allowed to live long enough distinctly to communicate the discovery. Nor could I resist the impulse — unphilosophical as it may have been — to cut with a knife on a stone of Brougham Bridge, as we passed it, the fundamental formula with the symbols $\underline{i}$, $\underline{j}$, $\underline{k}$; *namely,*

$$\underline{i}^2 = \underline{j}^2 = \underline{k}^2 = \underline{ijk} = -1,$$

which contains the solution of the problem, but of course as an inscription has long since mouldered away. A more durable notice remains, however, in the Council Book of the Academy for that day, which records the fact that I then asked for and obtained leave to read a paper on Quaternions, at the first general meeting of the session, which reading took place accordingly on Monday the 13th of November following."

Hamilton defined a quarternion as a number of the form $\alpha + \beta\underline{i} + \gamma\underline{j} + \delta\underline{k}$ where α, β, γ, δ are real numbers and $\underline{i}$, $\underline{j}$, $\underline{k}$ are three independent roots of -1. α is called the scalar part of the quaternion and the rest is called the vector part. We add quaternions componentwise and in multiplication note the rules

$$\underline{i}^2 = \underline{j}^2 = \underline{k}^2 = -1,\ \underline{ij} = -\underline{ji} = \underline{k},\ \underline{ki} = -\underline{ik} = \underline{j},\ \underline{jk} = -\underline{kj} = \underline{i}.$$

Thus if

$$p = 2 + 3\underline{i} + 4\underline{j} + 5k \qquad \text{and } q = 1 + 4\underline{i} + 2\underline{j} + 3\underline{k}$$

$$pq = 2 - 1\overline{2} - \overline{8} - 1\overline{5} + (8\underline{i} + 3\underline{i} + 12 - 1\overline{0})\underline{i}$$

$$(4 + 4 - 9 + 20)\underline{j}\,(6 + 5 + 6 - 16)\,\underline{k}$$

$$= -33 + 13\underline{i} + 19\underline{j} + \underline{k} \text{ which is another quaternion.}$$

whereas

$$qp = -33 + 9\underline{i} - 3\underline{j} + 21\underline{k}$$

Hamilton in 1844 proved that quaternion multiplication was associative (the first use of that term). He also introduced an important differential operator denoted by the symbol ∇. ∇ which is an inverted Δ he called "nabla", owing to its similarity to an ancient Hebrew musical instrument of that name. When applied to a scalar point function $u(x, y, z)$

$$\nabla u = \frac{\partial u}{\partial u}\underline{i} + \frac{\partial u}{\partial y}\underline{j} + \frac{\partial u}{\partial z}\underline{k}$$

This vector is now described as grad u.

If

$$V = V_1\underline{i} + V_2\underline{j} + V_3\underline{k}$$

denotes a vector function then

$$\nabla V = -\left(\frac{\partial V_1}{\partial x} + \frac{\partial V_2}{\partial y} + \frac{\partial V_3}{\partial z}\right) + \left(\frac{\partial V_3}{\partial y} - \frac{\partial V_2}{\partial z}\right)\underline{i} + \left(\frac{\partial V_1}{\partial z} - \frac{\partial V_3}{\partial x}\right)\underline{j} + \left(\frac{\partial V_2}{\partial x} - \frac{\partial V_3}{\partial y}\right)\underline{k}$$

The scalar part of this is now called the divergence of V (with a change of sign). The vector part is now called the curl of V.

Hamilton was to spend the rest of his life "selling" his quaternions. He believed they were important, not only for mathematics but also for physics. In this he was supported by P. G. Tait who in a succession of articles advocated quaternions to physicist. However, many initially found them too complicated and when later vectors were developed Quaternions were all but superseded.

It is one of the interesting aspects of the history of mathematics that almost at the same time as Hamilton was working out the vital part of his quaternions another non-commutative algebra was being developed by Grassmann.

In 1844 Hermann Günther Grassmann introduced his concept of hypernumber. Grassmann's hypernumber were essentially concerned with n-dimensional geometry and contained the germs of ideas that were to develop late into vectors and tensors. For simplicity we shall consider the case $n=3$. A hyper number α is of the form $\alpha = a_1e_1 + a_2e_2 + a_3e_3$ where α_1, α_2, α_3 are real numbers and e_1, e_2, e_3 are primary or qualitative units represented geometrically by directed line components. Grassmann considered two types of multiplication.

For inner products he postulated that $e_i | e_i = 1$, $e_i | e_j = 0$, $i \neq j$. For the outer product $[e_i, e_j] = -[e_j, e_i]\ [e_j, e_i] = 0$.

Where $[e_i, e_j]$ is called a unit of the second order and not reduced by Grassmann to primary units. However, Grassmann did tend to treat them as if they were equivalent to primary units with $[e_1, e_2]$ equivalent to e_3, etc. Thus we have the origins of the vector dot and vector cross product.

In a later paper (1855) Grassmann was to give for the n-dimensional hypernumber no less than 16 different types of product. In this paper he also quoted many applications to mechanics, magnetism and crystallography. However, for some time, mathematicians neglected Grassmann's work, unlike quaternions, they wanted something far more intimately related to three-dimensional cartesian space. A step in this direction was taken by J. Clark Maxwell in separating the scalar and vector parts of the quaternions.

The formal break with quaternions and the development of vectors as an entity in their own right, was made by two mathematicians in the early 1880s J. W. Gibbs at Yale and Oliver Heaviside began independently to develop a theory of vectors.

Gibbs was a mathematical physicist and as he wrote in a pamphlet for his students on *Vector Analysis* published in 1881 and 1884.[3]

"The fundamental principles of the following analysis are such as are familiar under a slightly different form to students of quaternions. The manner in which the subject is developed is somewhat different from that followed in treatises on quaternions, being simply to give a suitable notation for those relations between vectors, or between vectors and scalars, which seem most important, and which lend themselves most readily to analytical transformations, and to explain some of these transformations. As a precedent for such a departure from quaternionic usage Clifford's *Kinematics* may be cited. In this connection the name Grassmann may be also mentioned, to whose system the following method attaches itself in some respects more closely than to that of Hamilton."

The material developed in this pamphlet was extended and developed and finally published in 1901 as *Vector Analysis* by Gibbs and Wilson. The latter author based most of the text on lectures given by Gibbs.

Heaviside was a Post Office engineer who had studied quaternions but considered them far too hard for busy engineers to master. In a sequence of papers in *Electrician* during the 1880s he developed his own shorthand vector calculus. Vectors as we know them today derive from these two sources. Gibbs and Heaviside successfully fused together the ideas of Hamilton and Grassmann.

The word "matrix" was given its current mathematical definition by Sylvester in 1850. For many years up to that date mathematicians had studied determinants and their application to many areas of mathematics. The determinant of an array is a number and it was Sylvester who coined the word matrix when he wished to refer to the array of numbers as a whole. However, most of his subsequent work in this area was concerned with the evaluation of the determinant. It was Sylvester's friend the English mathematician Arthur Cayley who was the first to consider a theory of matrices independent of the need to evaluate a determinant. While it is true, as Cayley claimed in his 1850 paper that the idea of a matrix logically precedes that of a determinant the historical development has been the other way round. Thus, the basic properties, including row operations, etc, of matrices were already well known and understood by the time matrices were developed. Cayley himself acknowledged that determinants were one of the influences on his development of matrices. The other main influence was the convenience of matrices to express equations.

It was in connection with the study of invariants under linear transformations that Cayley first introduced matrices to simplify the notation involved. In 1855 in his paper for *Crelle's Journal*, "Fur Reine and Angewandte Mathematik", Cayley wrote:

$$(\xi, \eta, \theta \ldots) = \begin{vmatrix} \alpha & \beta & \gamma \\ \alpha' & \beta' & \gamma' \\ \alpha'' & \beta'' & \gamma'' \end{vmatrix} (x, y, z \ldots)$$

as shorthand for the set of equations

$$\begin{aligned} \epsilon &= \alpha x + \beta y + \gamma z + \ldots \\ \eta &= \alpha' x + \beta' y + \gamma' z + \ldots \\ \theta &= \alpha'' x + \beta'' y + \gamma'' z + \ldots \end{aligned}$$

The same paper also introduced the concepts of inverse matrix and matrix multiplication but neither idea was dealt with in any depth. Three years later in the Philosophical Transactions of the Royal Society of London, Cayley published a paper entitled "A memoir on the theory of Matrices" in which many of the basic properties of matrices were expounded. Here we find the definition of matrix equality the unit matrix and the zero matrix denoted by I. We find the definition of sum and scalar product of matrices. The definition of matrix multiplication was suggested to Cayley by the composition of two transformations.

If the transformation

$$\begin{aligned} f\colon x' &= a_{11}x + a_{12}y \\ y' &= a_{21}x + a_{22}y \end{aligned}$$

is followed by

$$g\colon x'' = b_{11}x' + b_{12}y'$$

The result is

$$\begin{aligned} x'' &= (b_{11}a_{11} + b_{12}a_{21})x + (b_{11}a_{12} + b_{12}a_{22})y \\ y'' &= (b_{21}a_{11} + b_{22}a_{21})x + (b_{21}a_{12} + b_{22}a_{22})y \end{aligned}$$

Hence Cayley defined the matrix product

$$\begin{pmatrix} b_{11} & b_{11} \\ b_{21} & b_{22} \end{pmatrix} \begin{pmatrix} a_{11} & a_{12} \\ a_{21} & a_{22} \end{pmatrix}$$

to be

$$\begin{pmatrix} b_{11}a_{11} + b_{12}a_{21} & b_{11}a_{12} + b_{12}a_{21} \\ b_{21}a_{11} + b_{22}a_{21} & b_{21}a_{12} + b_{22}a_{22} \end{pmatrix}$$

giving our present definition of matrix multiplication. Cayley also defined the inverse of

$$\begin{pmatrix} a & b & c \\ a' & b' & c' \\ a'' & b'' & c'' \end{pmatrix}$$

to be the matrix

$$\frac{1}{\nabla}\begin{pmatrix} \partial a\nabla & \partial b\nabla & \partial c\nabla \\ \partial a'\nabla & \partial b'\nabla & \partial c'\nabla \\ \partial a''\nabla & \partial b''\nabla & \partial c''\nabla \end{pmatrix}$$

where ∇ is the determinant

and $\partial_x\nabla$ is obtained from ∇ by replacing the elements in the row and column containing x by zeros. (This in current terms is called the cofactor of x.) Cayley also introduced the term indeterminate for the case $\nabla = 0$, and many other ideas like, transpose, symmetric, matrix, etc, were introduced. Cayley also gave a proof of the theorem now known as the Cayley-Hamilton Theorem.

In these two papers Cayley developed many of the concepts of matrix theory and the notation that he chose was so apt that much of it is preserved to this day. However, it is arguable how influential these papers were.

In a recent series of papers, T. Hawkins argues that there is more to a theory than its formal aspects. There is also its content which is equally if not more important. Hawkins argues that the origins of the content of matrix theory can be traced back to Gauss's work on determinants. The reader interested in this aspect of matrix theory is referred to Hawkins papers.

References

G. Flegg (1976) *Modern Algebra*, AM 289 Course Unit 9, The Open University Press.

J. W. Gibbs and E. B. Wilson (1901), *Vector Analysis*, Dover reprint (1960).

H. G. Grassmann (1844) *Die Lineale Auschlehnungslehre.*

Sir W. R. Hamilton *Elements of Quaternions*, 2 vols, 1866, Chelsea (reprint), 1960.

T. Hawkins (1975) Cauchy and the Spectral Theory of Matrices, *Historia Mathematica*, **2**, 1-29.

M. Kleine (1972) *Mathematical Thought from Ancient to Modern Times*, Oxford.

1. Quoted with permission of the publishers from: G. Flegg, *Modern Algebra*, AM 289 Course Unit 9, The Open University Press, 1976.
2. Quoted with permission of the publishers from: G. Flegg, op. cit.
3. Quoted with permission of the publishers from: *Mathematical Thought from Ancient to Modern Times*, by M. Kline, Oxford University Press, 1972.

Egyptian fractions in the classroom

by Paul Ernest, Homerton College, Cambridge

In teaching a top third-year set at Hampstead School last year I was faced with the problem of how to stretch "high flying" pupils working ahead of the majority of the set on SMP books F, G or X. I wished to broaden these pupils' mathematical background, in addition to helping them through "O" level early. My answer was to direct reading on the "History of Mathematics" and to follow up the readings with worksheets. The book I used was that by Popp[1], as much for its high ratio of mathematical concepts to rhetoric as for its cheapness and easy availability. The first worksheet is shown in Figure 1. (The numbering follows that of Popp.) The answers to this sheet were generally correct, but the most interesting answer to question 1.2.1 was wrong. The answer had the following form:

"1.2.1 (a) $\frac{5}{6}=\frac{3}{6}+\frac{2}{6}=\frac{1}{2}+\frac{1}{3}$

(b) $\frac{17}{20}=\frac{10}{20}+\frac{5}{20}+\frac{2}{20}=\frac{1}{2}+\frac{1}{4}+\frac{1}{10}$

(c) $\frac{23}{28}=\frac{14}{28}+\frac{7}{28}+\frac{2}{28}=\frac{1}{2}+\frac{1}{4}+\frac{1}{14}$

Not every fraction can be written as a sum of fractions with numerator 1. Fractions with a prime denominator, e.g. $\frac{3}{7}$, do not have enough factors in the denominator to add up to the numerator, as in examples (a), (b) and (c)."

The pupil's reasoning is incorrect for:

$$\frac{3}{7}=\frac{12}{28}=\frac{7}{28}+\frac{4}{28}+\frac{1}{28}=\frac{1}{4}+\frac{1}{7}+\frac{1}{28}$$

Far more interesting is the suggestion that a fraction with a denominator sufficiently rich in factors to total the numerator can be expanded into a unit fraction sum.

History of Mathematics Sheet 1

Book: "History of Mathematics" by W. Popp.
Read Chapter 1 and answer these questions as you finish each section.

1.1.1 (a) What numbers do these Cretan numerals represent?

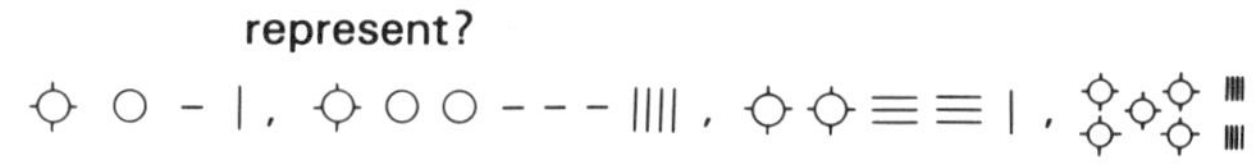

(b) What numbers do these Egyptian numerals represent?

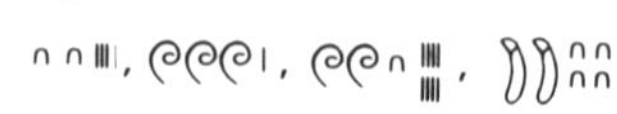

(c) Roman Numerals. What does the "C" mean in "CM"?
What is the value of CM?
What are the values of MMDCCCLXXXVII, MCMLXXIX?

(d) Greek Numerals. What numbers are represented by:

β, λ, $\phi\eta$, $\omega o\theta$, ͵τ͵π͵$\alpha\lambda\nu\epsilon$?

How many symbols do the Greeks need to show all the numbers from (1) 1 to 99; (2) 241 to 327?

1.1.3 Add MDCCLXVII and CCCXIII using abacus diagrams as on page 8. Give the answer both in Roman and Arabic (normal) numerals. (NB: D=CCCCC, L=XXXXX, V=IIIII).

1.1.4 Write the sum of the divisors (excluding the number itself) for each of the numbers from 20 to 30. Then classify these numbers according to whether they are *deficient*, *perfect* or *abundant* numbers.

1.2.1 The Egyptian method of writing the fraction $\frac{111}{160}$ uses the sum $\frac{1}{2}+\frac{1}{10}+\frac{1}{16}+\frac{1}{32}$. Write the following as sums of fractions with numerator 1: (a) $\frac{5}{6}$, (b) $\frac{17}{20}$, (c) $\frac{23}{28}$.
Do you think any fraction can be written like this? Explain.

Fig. 1

History of Mathematics Sheet 3

More on Egyption Fractions: **Sheet 1 ended by asking can all proper fractions be expressed in the Egyptian way as a sum of different fractions with numerator one?**

The answer is yes, and here is a way to do it applied to the example $\frac{10}{11}$:

Method	*Example*
1. Continue the sequence 1, 2, –, –, by repeated doubling. Stop when you cannot add any more terms less than your denominator.	1, 2, 4, <u>8</u>. ($8 < 11 \leqslant 16$)
2. Multiply up your fraction by the last term of the sequence to make an equivalent fraction.	$\frac{10}{11} \xrightarrow{\times 8} = \xrightarrow{\times 8} \frac{80}{88}$
3. Divide your new numerator by the original denominator.	$80 \div 11 = 7$ rem 3
4. Express your answer and remainder as sums of base 2 column headings.	BASE 2 COLUMNS (see table below)
5. Multiply each part of your answer by the original denominator and write as an addition sum with the remainder parts.	$4\times 11 + 2\times 11 + 1\times 11 + 2 + 1$ $= 44 + 22 + 11 + 2 + 1$
6. Put each part separately over the new denominator and simplify. (First check they add to your original fraction.)	$\frac{44}{88} + \frac{22}{88} + \frac{11}{88} + \frac{2}{88} + \frac{1}{88} (= \frac{80}{88})$ $= \frac{1}{2} + \frac{1}{4} + \frac{1}{8} + \frac{1}{44} + \frac{1}{88} = \frac{10}{11}$.

BASE 2 COLUMNS

	8	4	2	1	
ans 7 =		1	1	1	= 4 + 2 + 1
rem 3 =			1	1	= 2 + 1

Try this method on (a) $\frac{2}{3}$; (b) $\frac{4}{5}$; (c) $\frac{5}{9}$; (d) some fractions of your own. In each case show your working and check your answer.

Fig. 2

Flowchart for Sylvester's Decomposition

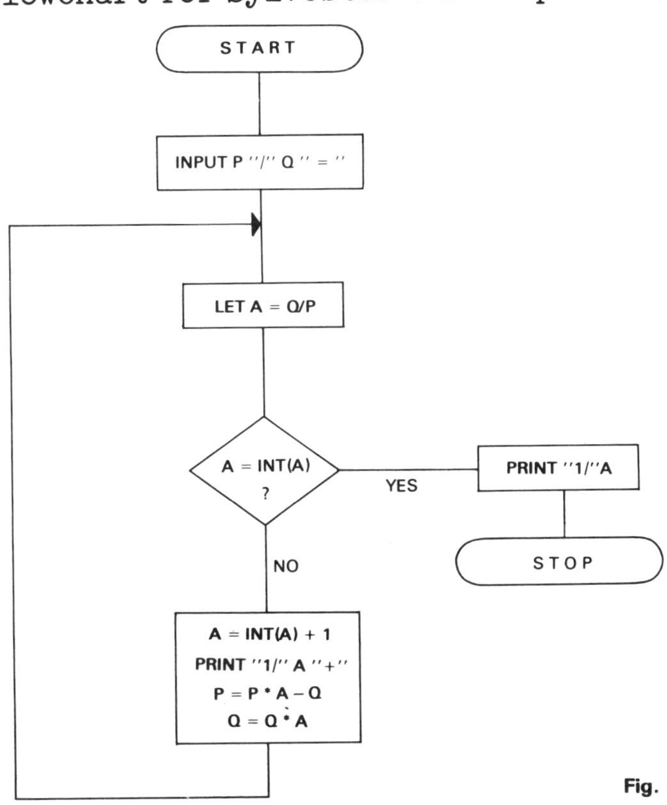

Fig. 3

Solving the Problem

Intrigued by this insight and the problem itself I set out to look for an equivalent $\frac{n.p}{n.q}$ of a proper fraction $\frac{p}{q}$ such that there are distinct factors of $n.q$ to sum to $n.p$. A general solution requires instead of n a function $f(q)$ such that $q.f(q)$ is sufficiently rich in distinct factors to sum to any of $2.f(q)$, $3.f(q)$, . . . , $(q-1).f(q)$, or for simplicity's sake, to any number less than $q.f(q)$. Then any $\frac{p}{q}$ in the form $\frac{p.f(q)}{q.f(q)}$ can be decomposed into unit fractions.

My first choice was $f(q) = (q-1)!$ However, whilst it is plausible that factors of $q! = q.(q-1)!$ can be chosen to sum to any number less than $q!$ the factors are rather complex. More importantly $(q-1)!$ is in general rather an inefficiently large choice for $f(q)$, as decomposing $\frac{3}{7}$ via $\frac{3.6!}{7!} = \frac{2\,160}{5\,040}$ suggests. Trying to find a suitable smaller value $m < q$ such that $f(q) = m!$ was rather daunting in view of the complexities of the factors involved, so I looked for a function whose factors are easily enumerated.

The choice $f(q) = 2^q$ gives a function with transparently clear factors. In this case $q.f(q) = q.2^q$ has factors including 1, 2, 4, . . . , 2^q, q, $2q$, $4q$, . . . , $2^{q-1}q$. To decompose $\frac{p}{q} = \frac{p2^q}{q2^q}$ into distinct unit fractions it is sufficient to express $p.2^q$ as a sum of distinct factors from the above list, for these factors put over $q.2^q$ simplify to distinct unit fractions. But it is easy to see that $p.2^q$ is a sum of these factors since $p.2^q$ can be written as a multiple of q and a remainder. The multiple of q can be expressed in terms of q, $2q$, $4q$, . . . , $2^{q-1}q$ which can express all multiples of q up to $(2^q - 1).q$. The remainder being less than q can be expressed as a sum in terms of 1, 2, 4, . . . , 2^q.

I made a final simplification in the choice of function by taking $f(q) = 2^n$, where 2^n is the greatest power of 2 less than q $(2^n < q \leqslant 2^{n+1})$.

I presented the algorithm[2] to the pupils in worksheet 3 (Fig. 2). I am not sure that I succeeded in verbally explaining *why* the algorithm works, although the pupils used the method successfully. The reading and worksheets continued until end of year exams stopped play. Retrospectively, I feel that the pupils gained experiences of elementary mathematics (for example, the addition of fractions) in the context of conceptually non-trivial problems, as well as a broadening of their historical/cultural perspectives of mathematics.

An Earlier Solution

After finishing with Egyptian fractions, I discovered that the same problem had been formulated and solved by Sylvester in the nineteenth century[3]. Sylvester's approach is different. He subtracts the largest unit fraction not exceeding his starting fractions, and he reiterates the process with the remainder. Sylvester's algorithm is shown in the flow chart (Fig. 3) Sylvester's proof that this procedure terminates is as follows:

If $\frac{1}{k}$ is to be subtracted from $\frac{n}{m}$, then $\frac{1}{k-1} > \frac{n}{m} \geqslant \frac{1}{k}$.

Subtracting: $\frac{n}{m} - \frac{1}{k} = \frac{nk-m}{mk}$, but by the first inequality $n > kn - m$. That is, the numerator of the remainder is a natural number less than the numerator of the original fraction. Hence the iterated procedure must terminate in at most n steps.

References

1. Popp, W. (Translated by M. Bruckheimer) (1975) *History of Mathematics — Topics for Schools* Transworld Student Library, Transworld Publishers Ltd.
2. A flow chart for the Algorithm in Figure 2, a rigorous proof that it works and extensions of this result are given in: Ernest, P. (In Press) "On the adequacy of the Egyptian representation of fractions" *Bulletin of the Institute of Mathematics and its applications*.
3. Bunt, L. N. H., Jones, P. S. and Bedient, J. D. (1976) *The Historical Roots of Elementary Mathematics* Prentice Hall. For Sylvester's proof see the exercises following Chapter 1 and Hints and Answers.

History as it is wrote!

by Frank Tapson, Clyst Vale Community College, Devon

The setting of a project is easy, but when the projects are passed in 6 weeks later how many of us wish that we hadn't started it? Recently I initiated some work on the history of mathematics and the heavy slog of marking was enlivened only by the misquotes and evident misunderstandings. I kept a note of them as I went along and here they are, completely unaltered—my notes in parentheses are only there to make clear the subject matter. Apart from the humour (nearly always unintentional) that they provide, I think they say a lot about the information and impressions gained by the children from their reading—which was mainly guided by me.

"Primitive man couldn't count. First man began to count sheep and pebbles. Then he began corresponding one to one"

"We know, because of history, that primitive man couldn't count"

"Primitive man couldn't count. He didn't really need to count. Unfortunately it didn't stop there"

(on the Babylonians) "It is said that these people are to blame for the 60 seconds in one minute"

"Thales is said to be the first human being who could be called a man in science"

"Thales was used for finding the height of the great pyramid. This was in Egypt"

(on Thales) "He made no very relevant discoveries but he calculated that there would be a good olives harvest one year, and he bought up all the olive mills around his home and he made a good prophet"

(on Euclid) "He published the second best selling book in the world, on Geometry"

"Euclid wrote the world's best seller besides the bible. He was good on geometry"

"Euclid said, 'There is known royal road to geometry' "

"Pythagoras lived in Italy before christ"

"Pythagoras was one of a whole race of brothers called the pythagorians. These brothers took laws of secrecy, but one of them let out the secret of Rational numbers e.g.

$$1,\ 2,\ \tfrac{5}{2},\ \tfrac{7}{4},\ \tfrac{119}{52}$$

He said they were not rational, no one could understand this"

"He had a band called the Pythagoreans, these were mathematicians"

"I'm sure you've heard of the story of Archimedes running through the streets, naked (don't look Ethel) after discovering specific gravity in his bath." (He was obviously the first streaker)

(on Archimedes) "he said π was equal to $3\frac{1}{7} < \pi < 3\frac{10}{71}$"

(on Archimedes) "When he was 75 he had his head cut off"

"Archimedes discovered the law of specific gravity at about the same time as Newton. In fact they argued who had really discovered it first. One thing he was sure on was that he discovered the law of the lever"

"Napier home country was Scotland. He had a habit of getting pigeons drunk. He also invented logs"

(on Napier) "a grand scot who spent his life discovering logarithms"

(on Newton) "He was made game keeper of the mint"

"Newton went to Cambridge. There he met a girl and later got engaged to her but he was so envolved with his mathematical work that he had no time for her and the engagement was called off!!"

(on Newton) "At 19 yrs, he went to Cambridge and fell in love for the first time, but his work in mathematics was too important and he ignored this new 'faze' "

"Here he mat a girl to whom he became engaged, but Newton became so devoted to his work that his engagement had to be called off!"

"Calculus has been in existence for thousands of years but in 1642-1727 Newton, Leibniz invented the use of calculus in mathematics"

"Leibniz was a German scholar. He and Newton developed the calculus at the small time"

(on Fermat) "He said 'you won't find any other numbers that add like this—

$$3^2+4^2 = 5^2$$

and you won't ether!"

"The Frenchman, Fermat was know for one thing, it is—

$$3^2+4^2 = 5^2$$

It has troubled mathematicians ever sinse."

(on Gauss) "He was 3 years old when he corrected his fathers paper bill"

"The value of π hasn't changed very much in 3,000 years"

"Pi will never be solved but many people have tried"

"One mathematician proved that you could add, divide, multiply and subtract algebra" ■

6 Recreational Mathematics
by David Neal

Recreational mathematics has a long history and has developed through many civilisations because of the interest and fascination it has invoked. This small collection of articles taken from past editions of *Mathematics in School* has been chosen so as to illustrate how the journal has attempted to provide ideas which may be of interest in the classroom.

Many readers will be familiar with the popular section of recreational mathematics under the title *Puzzles, pastimes and problems*. However, some may be unaware of the extent of this collection which originated in the first edition of the journal in 1971. Its author, Canon D. B. Eperson, has developed many fascinating exercises within the series, a small sample of which have been chosen to introduce this selection.

The current picture of the learning of mathematics in school which emphasises activity, investigation, discussion and similar processes, has increased the incidence of recreational mathematics in school and this has increased the number of articles related to recreational mathematics which have been published in *Mathematics in School.* A recent, regular feature, *Maths resource*, developed by Frank Tapson, has proved to be a particularly popular supply of classroom-ready resource material which frequently develops recreational mathematics. A single example entitled *Dissections* has been chosen to illustrate the nature of the series.

A second regular feature, *Mathes games workshop*, developed by David Kirkby, has demonstrated how mathematical skills and concepts can be acquired through recreational activities. *Railway Rivals* by Alan Parr illustrates the understanding and fascination which this kind of recreational mathematics can produce in the classroom. Keith Porteous, in his article on mathematics competitions demonstrates the enjoyment-orientated competition involving the majority of pupils. 'A grid game' shows how a simple game, suitable for top juniors can be developed to sixth form level and at the same time provide a creative investigation with varied mathematics. Finally the investigation involving placing tiles on a grid illustrates a type of recreational activity which is a regular feature in the pages of *Mathematics in School.*

The popularity of recreational mathematics was perhaps most clearly illustrated in the January 1986 edition of *Mathematics in School* which devoted all its pages to this topic. This selection is complementary to that edition.

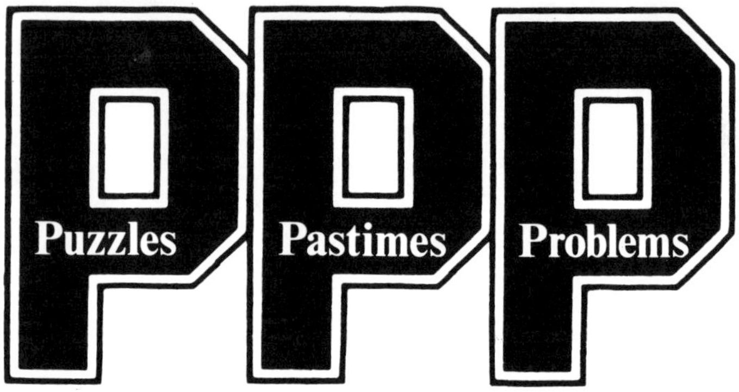

by D. B. Eperson

1. Symmetrical Patterns

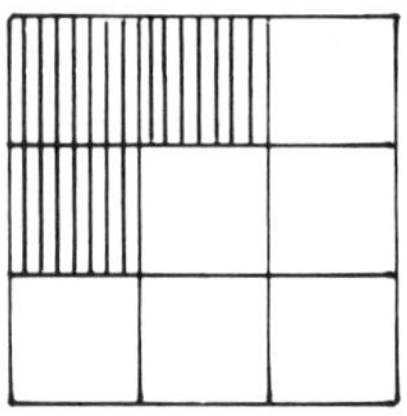

Fig. 1.1

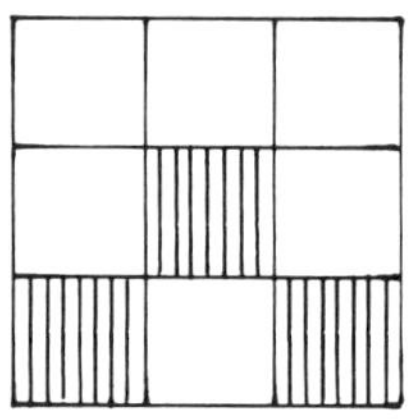

Fig. 1.2

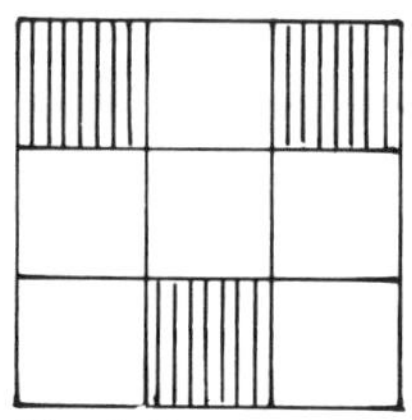

Fig. 1.3

1.1 Each of the above patterns on a 3×3 square has 3 black squares and 6 white squares, and a single line of symmetry:
   (a) How many more essentially different patterns of this kind with one line of symmetry can you find?
   (b) Can you find any patterns with 3 black squares and two lines of symmetry?

1.2 How many patterns can you find, made with 4 black squares and 5 white squares:
   (a) with one line of symmetry;
   (b) with two or more lines of symmetry?

Figure 1

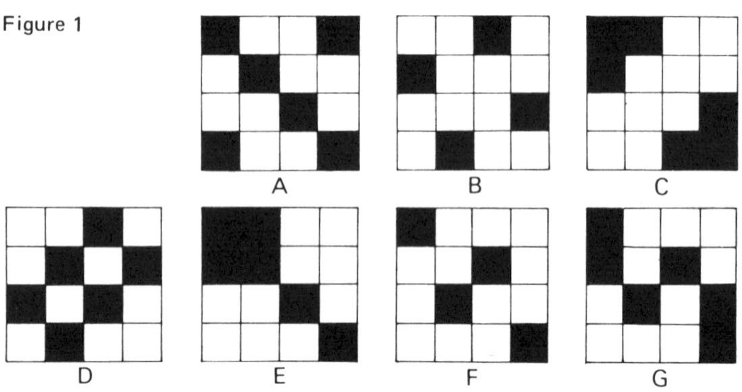

1. Symmetrical Number Patterns

(a) What numbers are represented by these black and white patterns?
(b) How many lines of symmetry has each pattern?
(c) Which of the patterns have rotational symmetry?
(d) Can you make some new patterns for these numbers which have
  (i) one line of symmetry only?
  (ii) two lines of symmetry only?
  (iii) four lines of symmetry only?
  (iv) rotational symmetry, but no line of symmetry?

1. Domino Patterns

A simple, but intriguing, pastime consists in making symmetrical patterns with dominoes that fill squares. The domino shape is a rectangle whose length is twice its breadth: you can use the plain backs of a set of 28 dominoes to make the patterns that fill squares, from 2 by 2 to 7 by 7, but for experimenting with the 8 by 8 "chess board" square, it is easy to cut up a square of cardboard into 32 dominoes by dividing it into halves, quarters, eighths, sixteenths and thirty-seconds in succession.

Fig. 1

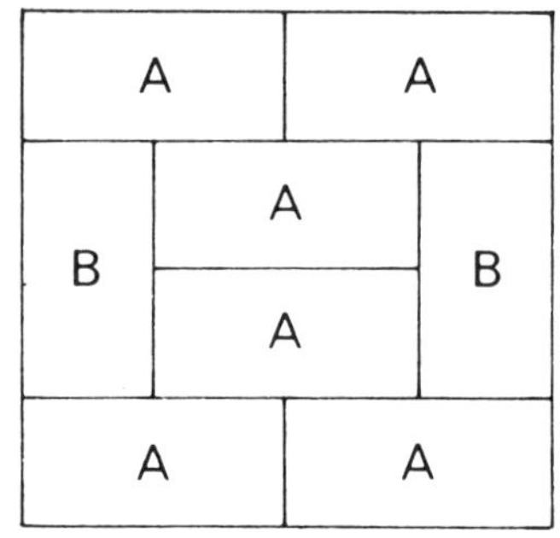

Fig. 2

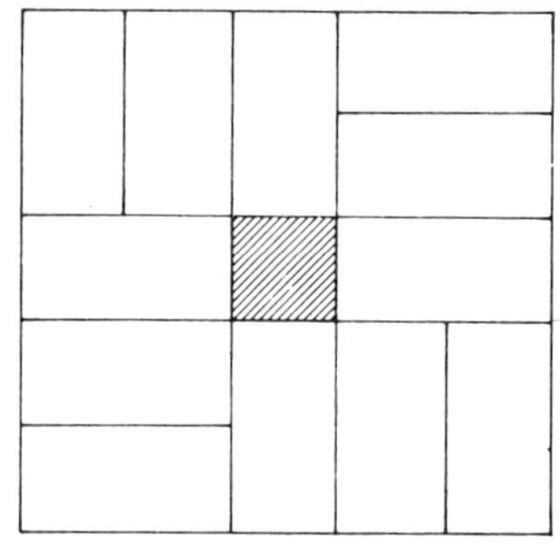

Fig. 3

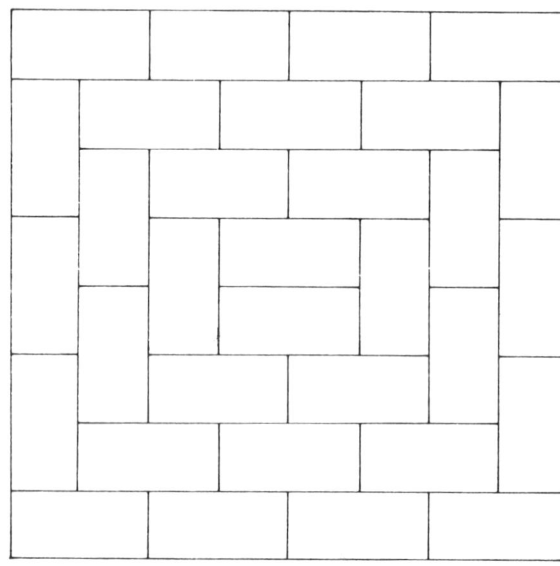

(a) Fig. 1 shows 8 dominoes filling a 4 by 4 square, making a pattern with two lines of symmetry. Six dominoes are in position A and two in position B. Can you find patterns with four dominoes in position A and four in position B, which have (i) two lines of symmetry, (ii) only rotational symmetry?

(b) Fig. 2 shows how to fill a 5 by 5 square, leaving the central cell empty, and making a pattern which has rotational symmetry only. Can you find patterns with rotational symmetry only to fill (i) a 3 by 3 square, (ii) a 5 by 5 square, (iii) a 7 by 7 square, leaving the central cell empty? (iv) Is it possible to find patterns with two lines of symmetry for these squares?

(c) Fig. 3 shows how to fill an 8 by 8 square, making a pattern with two lines of symmetry: 20 dominoes are in position A, and 12 in position B. Can you make patterns with 16 dominoes in each position to make patterns (i) with two lines of symmetry, (ii) with rotational symmetry only?
(d) Can you fill a 6 by 6 square to show patterns with each kind of symmetry?

Pastimes With Dominoes

When I was young I often amused myself after playing a game of dominoes by using them as building bricks, or as rectangular "tiles" to make patterns. Later I realised that I had been experimenting with tesselations. After 70 years the topic is still of recreational interest, and surprises reward the investigator.

4.1 Can you build a single "colonnade" (Figure 4.1) using all the dominoes?

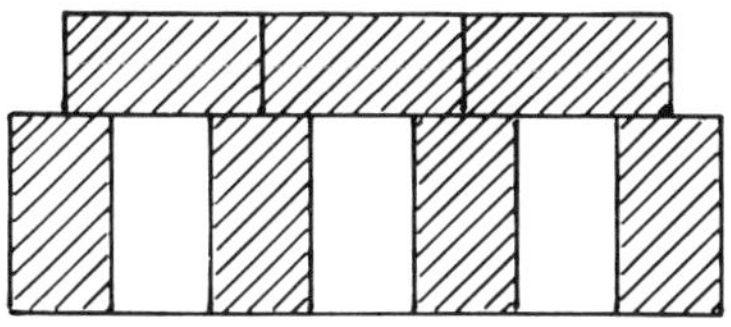

Fig. 4.1

What is the longest colonnade that can be built, and how many openings between pairs of pillars has it?

4.2 Can you build a double colonnade (Figure 4.2) using all the dominoes?

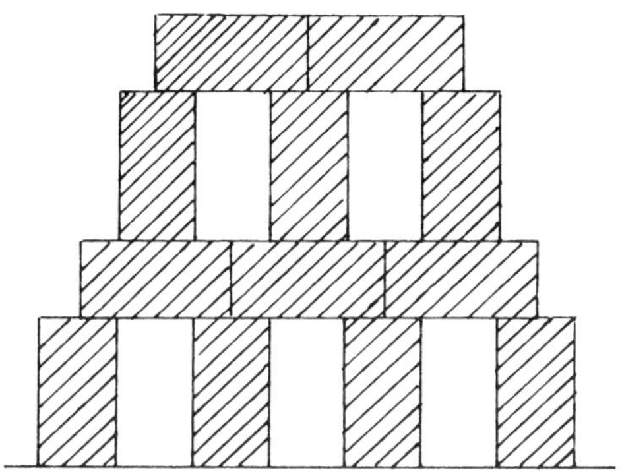

Fig. 4.2

4.3 Can you build a "triangular wall' (Figure 4.3) using all the dominoes?

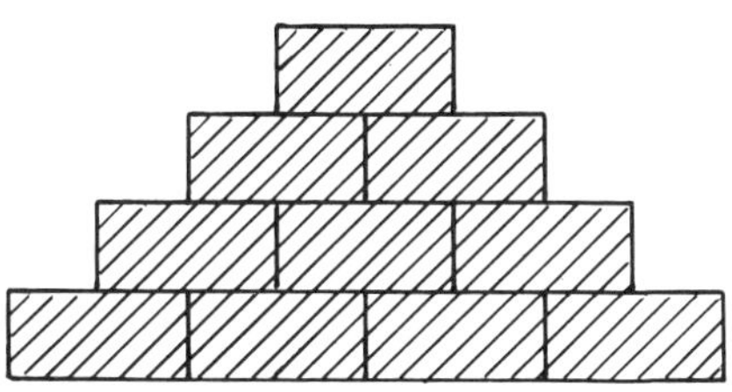

Fig. 4.3

4.4 Can you design a different style of single colonnade that uses all the dominoes?

4.5 Dominoes are usually rectangular, with length twice the breadth. Figure 4.5 shows three dominoes forming a 2×3 rectangle with one line of symmetry, that could be used as a unit when making an extensive tesselation. Use eight dominoes to make a square pattern with two lines of symmetry.

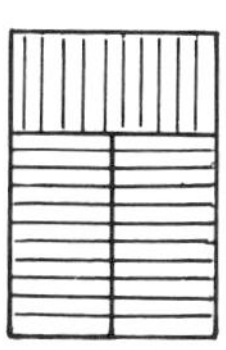

Fig. 4.5

4.6 Use eight dominoes to make a pattern with rotational symmetry.

4.7 Use all the dominoes to make a rectangular pattern with sides in the ratio 7:8 and two lines of symmetry.

4.8 Invent some patterns of your own.

1. Magic Square Jigsaw Puzzles

(i) A Chinese Magic Square, made with the numbers from 1 to 9, has been cut into the three pieces shown in Fig. 1. Can you fit them together to make the Magic Square?

(ii) Another Magic Square has been cut into two rectangles and two squares, shown in Figure 2: can you solve this Magic Square Jigsaw Puzzle?

Figure 1

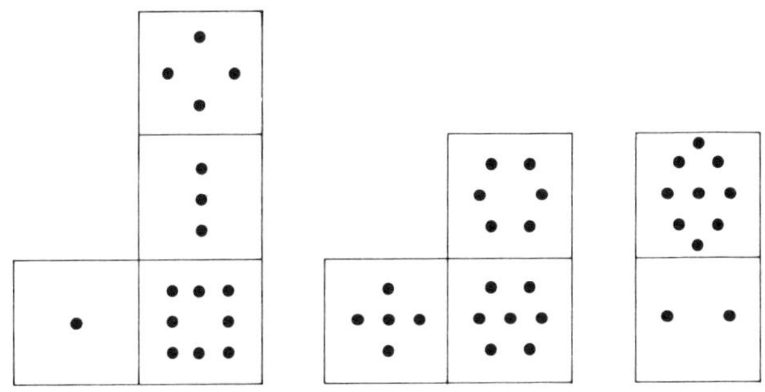

Figure 2

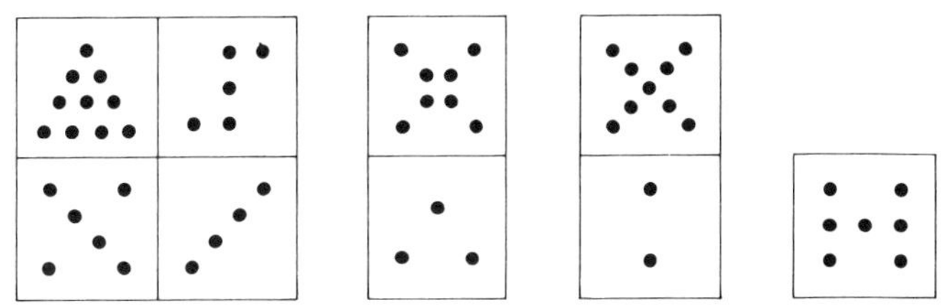

4. Magic Domino Squares

Four dominoes make a "Magic Domino Square" (in short MDS) if, when arranged to form a hollow square, as shown in Fig. 5, the sum of the numbers along each side of the square have the same total, T. In Fig. 5 we see that 3+5 = 6+2 = 2+1+5 = 8.

(a) Complete the MDS, Fig. 6, so that (i) T = 8, (ii) T = 9,

Figures 5 6 7 8

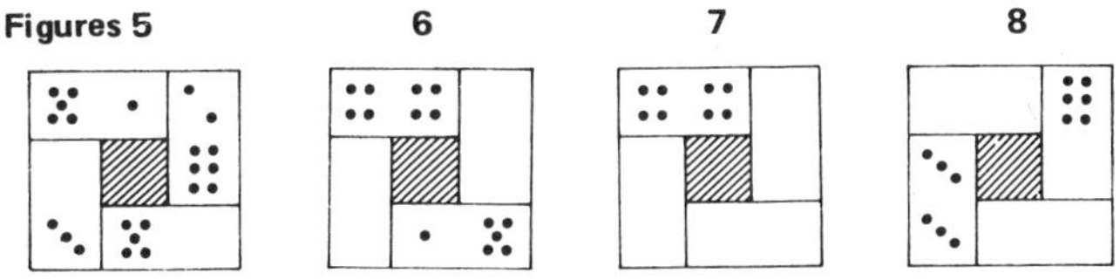

(iii) T = 10, (iv) T is the largest number possible. (Note: all the dominoes in a set are different, and show all pairs of numbers from 0 (blank) to 6, including "doubles".)

(b) Complete the MDS, Fig. 7, so that it contains two other doubles besides the double-four shown, with T = 8. Is it possible to do this when T has any other values?

(c) Complete the MDS, Fig. 8, so that T = 6.

(d) Make a set of Magic Domino Squares with T ranging from 4 to 14, without using any doubles.

(e) Construct the MDS with the smallest possible value for T.

(f) Construct the MDS with the largest possible value for T.

(g) Is it possible to make MDS with the same set of three figures along (i) all four sides, (ii) three of the sides?

(h) Show that a few MDS can be made with three doubles in them.

(i) Construct some MDS with two doubles (i) on adjoining sides, (ii) on opposite sides of the square.

Magic Squares From a Calendar

Select a square block of nine numbers from any calendar for a month, and you will find that it has some "magic" in it, like an ordinary Magic Square. Verify these results in the one

SUN	MON	TUES	WED	THUR	FRI	SAT
			1	2	3	4
5	6	7	8	9	10	11
12	13	14	15	16	17	18
19	20	21	22	23	24	25
26	27	28	29	30	31	

shown above: (a) the sum of the three numbers in each of the diagonals is the same as the sum of those in the central column; (b) these three sets of numbers form Arithmetic Sequences, with common differences of 6, 7 or 8; (c) the sums of the numbers in each of the rows also form an Arithmetic Sequence, with a common difference of 21; (d) the sums of the numbers in each of the three columns form an Arithmetic Sequence with a common difference of 3.

Now select a square block of your own, and verify that these relations (a), (b), (c), (d) hold for its numbers, and then prove that they hold for any block in the calendar.

Show also that in every case the nine numbers can be rearranged to make an ordinary Magic Square.

Finally show that any square block of 16 numbers from a calendar can be rearranged to make an ordinary 4×4 Magic Square.

A New Kind of Magic Square

Can you fit the numbers from 1 to 8 into the above square, so that all these sets of four numbers have the same total?

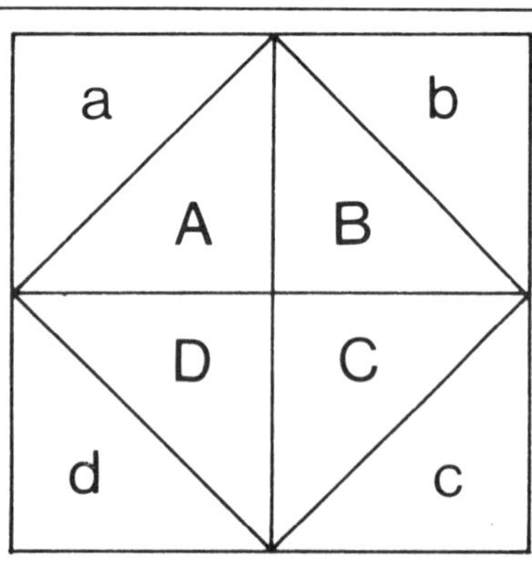

$a + b + c + d$, $A + B + C + D$, $a + b + A + B$, $a + c + A + C$, $a + d + A + D$, $b + c + B + C$, $b + d + B + D$, $c + d + C + D$.

In your solution, is it true that the sum of any pair of numbers represented by small letters is the same as the sum of the pair of numbers represented by the other pair of capital letters? (For example, does $a + c = B + D$, etc.?)

Sequences

1	3	5	7	9	11	13	15	17	19	21
1	4	7	10	13	16	19	22	25	28	31
1	5	9	13	17	21	25	29	33	37	41
1	6	11	16	21	26	31	36	41	46	51
1	7	13	19	25	31	37	43	49	55	61
1	8	15	22	29	36	43	50	57	64	71

Figure 1

The chart shows in the top row the sequence of odd numbers as far as 21: the second row is the sequence formed by adding 3 at each stage subsequent rows contain other arithmetic sequences based on the additive principle.

(a) Find the totals of the numbers in each row and column, and hence find the sum of all the numbers in the chart in two ways.

(b) Find formulae for the nth term in each row, and the mth term in each column.

(c) Where do all the square numbers appear in the chart?

(d) If the chart were extended by adding more columns on the right, in how many rows would each of these prime numbers appear? (i) 29, (ii) 31, (iii) 37, (iv) 43, (v) 61, (vi) 101.

(e) Draw the pattern formed by blacking out all multiples of 3. Explain why some rows contain no multiples of 3. Is it true that the black squares in the pattern are linked by Knight's moves on the chessboard?

(f) Which rows in the extended chart will contain numbers that are multiples of 10?

(g) The number 16 appears in the fourth row and the fourth column: find a formula for all the numbers that would appear in this row and this column, if the chart were extended both to the right and downwards.

Fun with Numbers

(i) Yes: it is always true. For example, if a six figure number has the digits a, b, c, d, e, f, it represents the number $100{,}000a+10{,}000b+1{,}000c+100d+10e+f$, and when this is added to the number with these digits in reverse order $100{,}000f+10{,}000e+1{,}000d+100c+10b+a$, the result is $100{,}001a+10{,}010b+1{,}100c+1{,}100d+10{,}010e+100{,}001f$, which is obviously a multiple of 11.

(ii) Yes: the difference is always a multiple of 11. For example, if a five figure number has the digits a, b, c, d, e, it represents the number $10{,}000a+1{,}000b+100c+10d+e$ and the difference between it and $10{,}000e+1{,}000d+100c+10b+a$ is $9999(a-e)+990(b-d)$, which is always a multiple of 11.

(iii) (a) Yes. (b) No. For example a palindromic number with six figures could be $100{,}000a+10{,}000b+1{,}000c+100c+10b+a$, which is $100{,}001a+10{,}010b+1{,}100c$, which is a multiple of 11 for all values of a, b, c. But a palindromic number of five figures, $10{,}000a+1{,}000b+100c+10b+a$ leaves a "remainder" $2a-2b+c$ when divided by 11, and this "remainder" will vanish only for a special set of values for a, b, c.

(iv) Using the previous answer, the palindromic number is a multiple of 11 if $2(X-Y)+Z = 0$, or 11, or 22, so that $Z = 2(Y-X)$, or $11-2(X-Y)$, or $22-2(X-Y)$. If $Z = 0$, then $X = Y$. 51315, 57475, 92829.

Palindromic Numbers

Consider this set of palindromic numbers:
1, 121, 12321, 1234321, 123454321, ... 12345678987654321.

(i) Find the set of numbers obtained by adding the figures in each number.

(ii) Find the set of remainders when each of these palindromic numbers is divided by (a) 2, (b) 4, (c) 5, (d) 8.

(iii) Find the patterns of the remainders after division by (a) 3, (b) 6, (c) 9, (d) 11.

A Knight's Tour on a Chess-Board

In Rouse Ball's *Mathematical Recreations and Problems* the author discusses the problem of "moving a Knight on a chessboard in such a manner that it moves successively to every cell once and once only". Eminent mathematicians such as Euler and Demoivre have shown how to obtain solutions on a full-size chessboard, but here are some simpler problems on a five-by-five mini-chessboard that anyone can tackle.

	12	7	2	
6	1		13	8
11				3
	5		9	
	10		4	

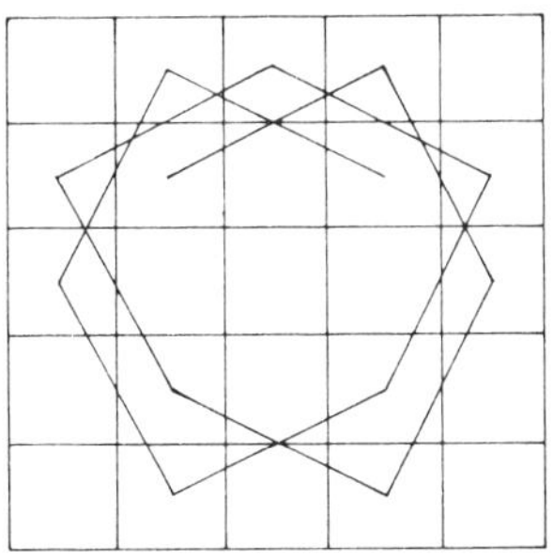

Figures 1, 2, 3

The numbered chart, Fig. 1, shows how to make a short tour visiting 13 cells numbered 1 to 13, so that the black and white pattern Fig. 2 made by the visited and unvisited cells has a line of symmetry, the tour diagram being shown in Fig. 3. Note that the middle cell of the tour, No. 7, is on the line of symmetry of the pattern. A tour can obviously be numbered in reversed order.

(a) Find the numbered chart and the tour diagram of the symmetrical pattern shown in Fig. 4.

(b) Show that there are two different numbered charts and tour diagrams for the pattern in Fig. 5.

(c) The pattern in Fig. 6 has two lines of symmetry, and rotational symmetry about the central cell. How many different numbered charts and tour diagrams can you find for Fig. 6?

(d) Invent your own symmetrical Knight's tours of this kind, and send some of them, with your solutions, to Canon Eperson, who wishes to publish more of these problems in future issues.

Figures 4, 5, 6

More Knight's Tours

In the last issue of *Mathematics in School* we found some Knight's tours on a mini-chess board that made black and white patterns with a line of symmetry. Now we investigate tours that visit 13 of the 25 cells, and make patterns with rotational symmetry, as shown below: Fig. 1 is the Number Chart, Fig. 2 is the Tour Diagram, and Fig. 3 is the Pattern formed by the 13 cells visited on the tour. Note that Fig. 1 has numerical symmetry, since pairs of numbers whose sum is 14 are in cells symmetrically placed with respect to the central cell, and that Figs. 2 and 3 both have rotational symmetry.

Figure 1

	6	1		11
		10	5	
	2	7	12	
	9	4		
3		13	8	

Figure 2

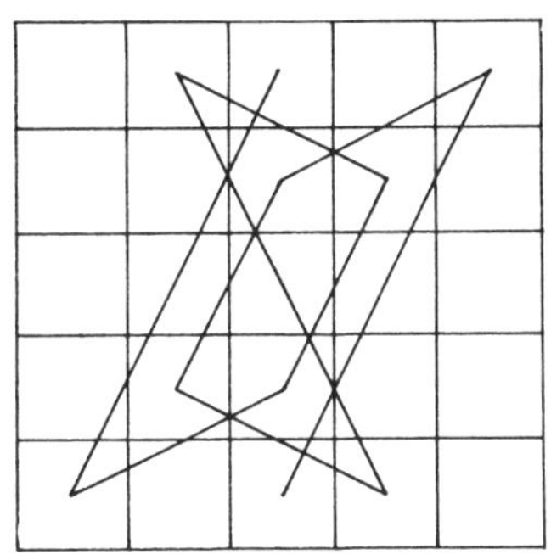

Figure 3

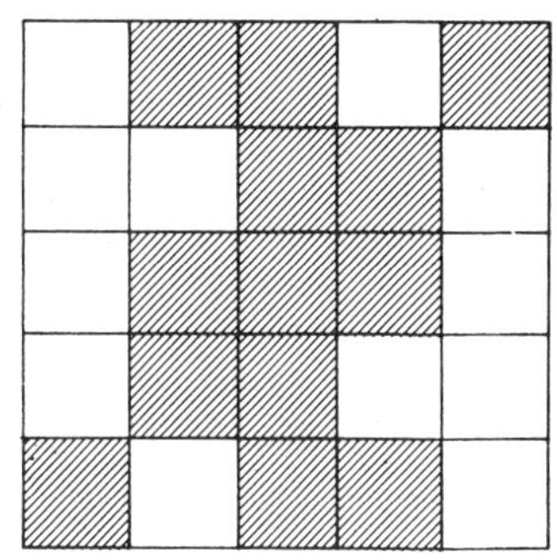

Find all the different number charts and tour diagrams that correspond to the patterns shown in Figs. 4, 5 and 6, which have rotational symmetry. No. 6 also has two lines of symmetry, but I cannot find any tour diagram that has a line of symmetry!

Figure 4

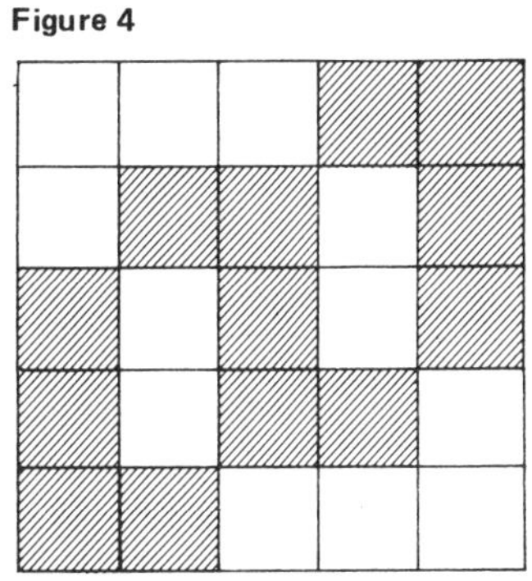

Figure 5

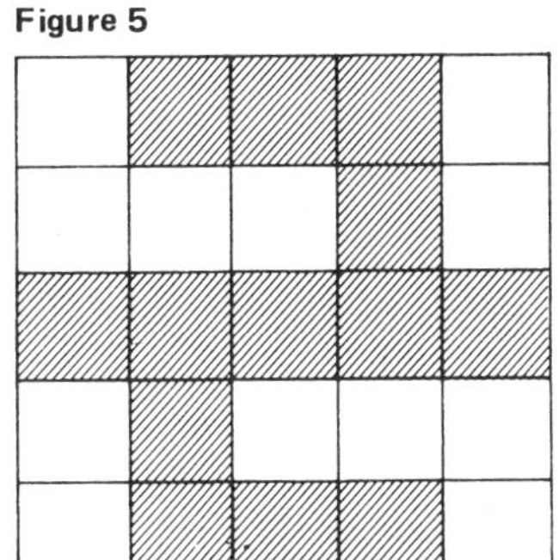

Figure 6

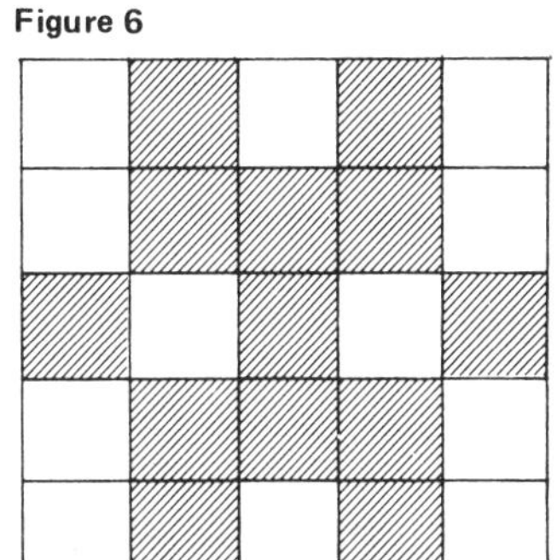

DISSECTIONS

by Frank Tapson

Each of the shapes shown here can be cut up, along the dotted lines, and the pieces re-arranged to make —

(a) a square
(b) a parallelogram
(c) an irregular quadrilateral
(d) a right-angled triangle

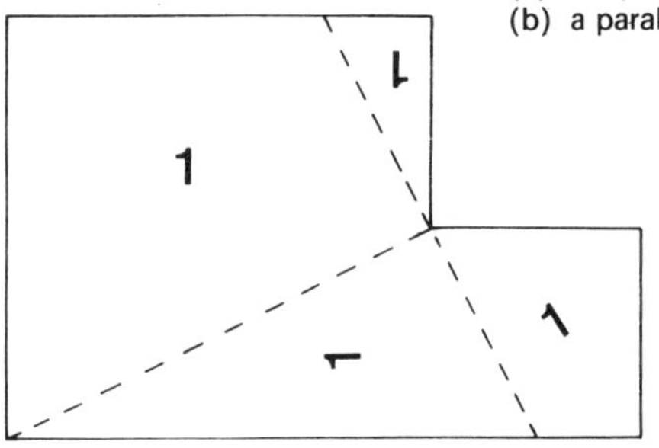
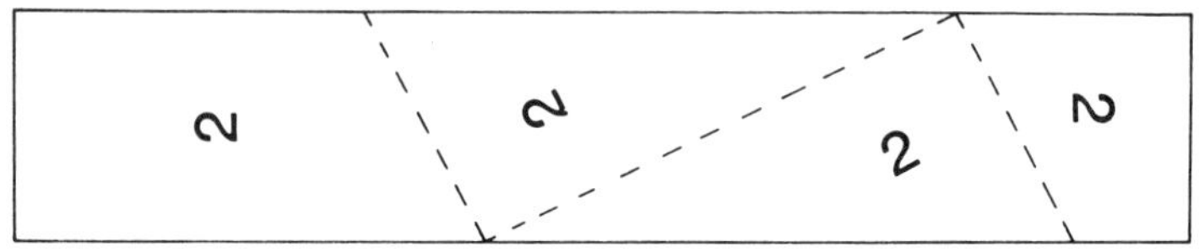
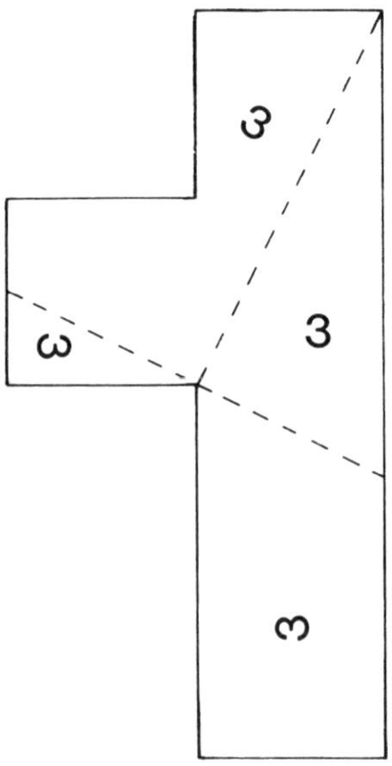
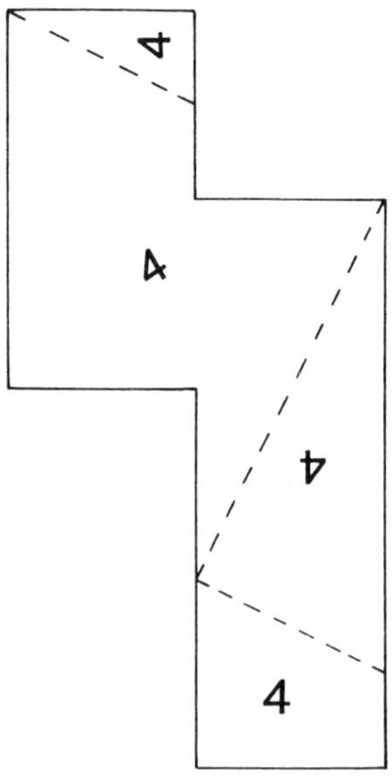
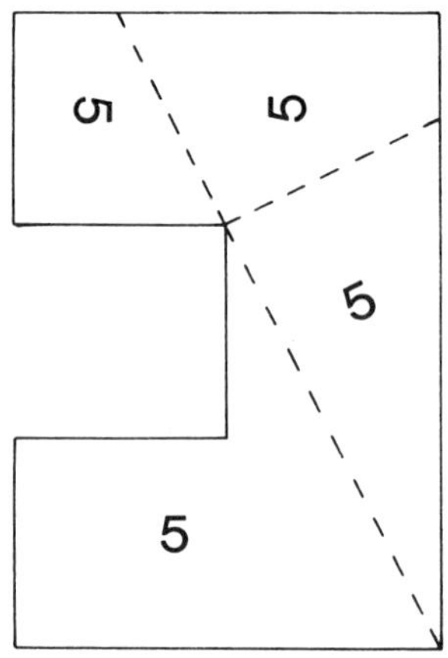
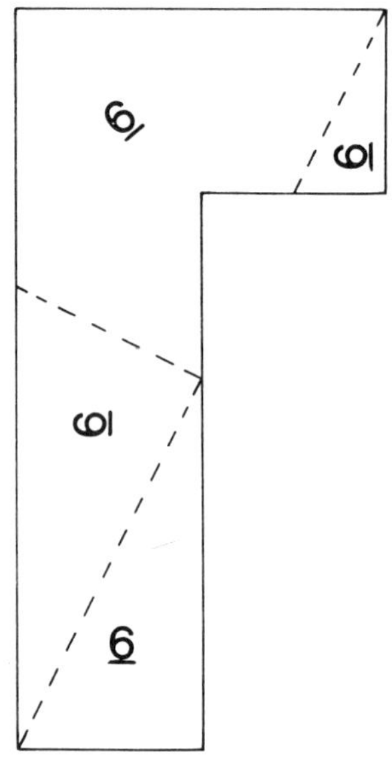
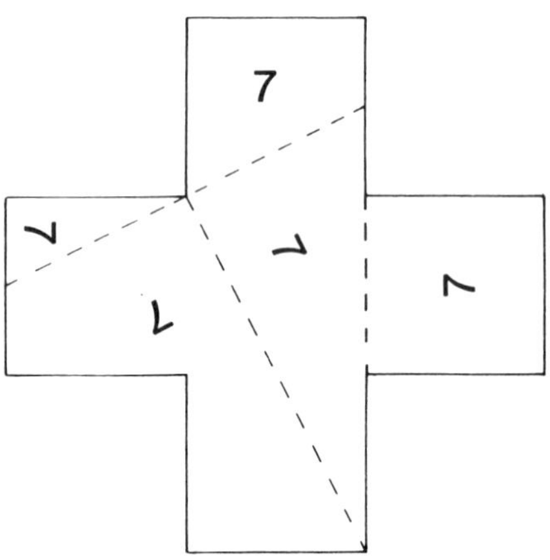
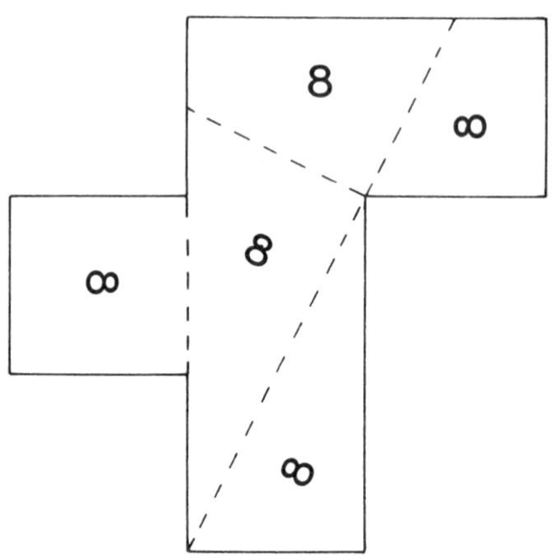
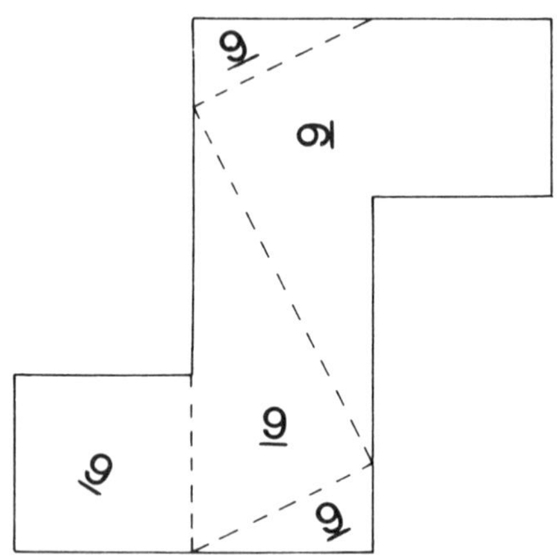
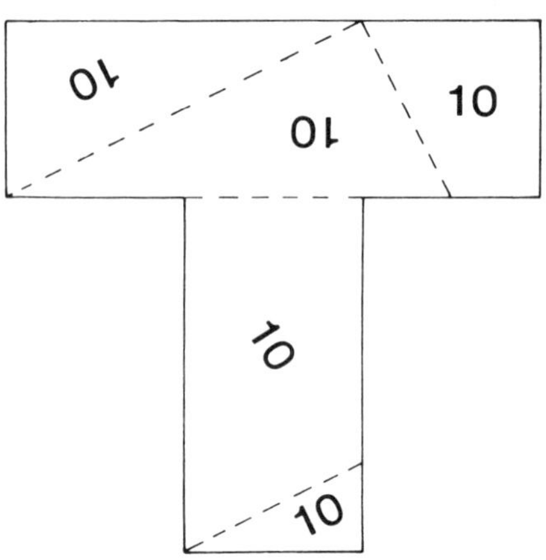
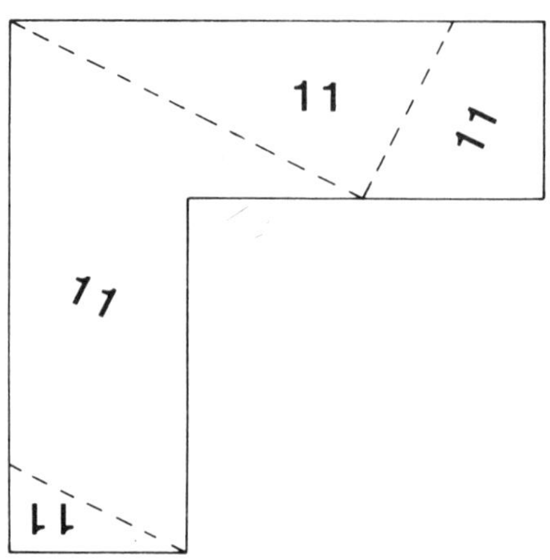
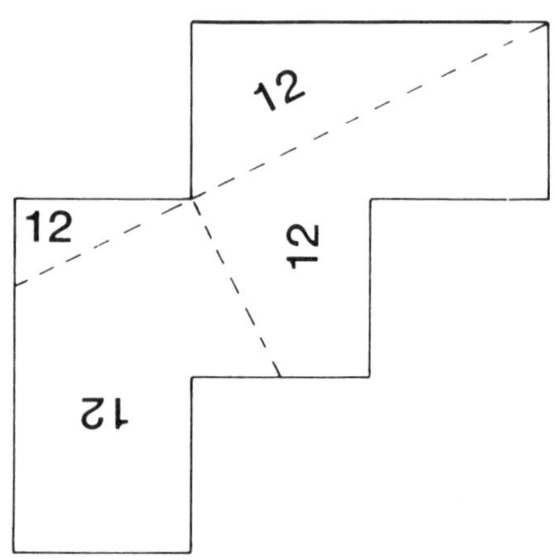

MRes 21 ©

RAILWAY RIVALS

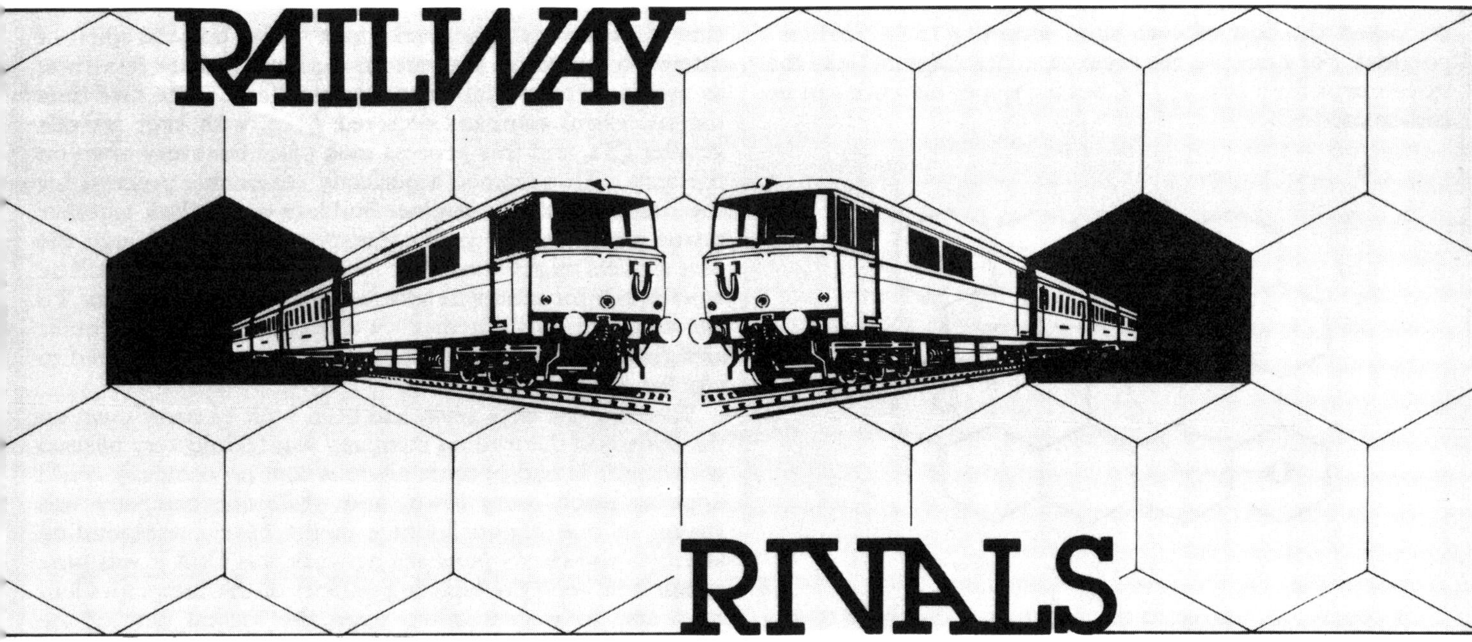

by Alan Parr, The Augustus Smith Middle School, Berkhamsted

It is now some ten years or more since a series of chance events led me to discover the revolution that was sweeping through geography teaching, and particularly the role found for the games and simulations being developed by J. P. Cole, Rex Walford, and others. Such games have over the years brought much understanding and enlightenment to my pupils and even myself (for many such games also exist in versions for adults to play by post — an activity which takes up all too much of my spare time).

Keith Selkirk's recent series of informative articles in *Mathematics in School* have highlighted the way in which such models have their basis in mathematics, and reminded me that some eighteen months ago I used one such game, "Railway Rivals" (the invention of another pioneer, David Watts), as the core of an extended mathematics topic for middle school children. This work progressed so well, and threw out so many interesting points that I subsequently wrote about it in a leaflet for interested teachers. What follows is the text — virtually unchanged — of what I wrote.

A Topic For Mathematics Lessons

This is an account of a topic I did recently with a group of 12-year olds. The work was in fact done in mathematics lessons, though if our school were structured differently I might have preferred the chance to use it as an integrated topic across the whole curriculum. It was a long topic and perhaps some people may be surprised both by its theme and by the way in which I based an extended period on a game — or rather a simulation. In choosing the topic, however, I was aware that the children were of well below average ability and that they had experienced only a fairly limited range of mathematics in which they had built up a large total of repeated failure, so I particularly wanted to show them that mathematics means more than lists of sums. I wanted, in short, to help them realise both that mathematics lessons can be enjoyable and that mathematics is a tool for operating in the world rather than a sterile set of exercises divorced from all reality. Furthermore, it goes without saying that we would actually be covering a good deal of mathematics almost without the children realising it — and this did indeed prove to be the case.

We had already done some work on towns, bus routes, timetables, and the like, and I decided to base the work upon an educational game called "Railway Rivals" (for full details of the availability of this, see Appendix). One contributory reason for the choice was that I'd read a review in an educational journal which implied that the game was only suitable for the brighter sixth-former; I'd known the game for years and knew this to be nonsense — indeed my own eight-year old was at that very moment playing in a postal game. Therefore I had no qualms about using the game with young children of less than outstanding ability.

Now "Railway Rivals" was designed as an aid to the teaching of geography, and it certainly possesses enough geographical, historical, and cultural merits to justify its use in the classroom, but it also seemed to promise enough mathematical benefits for it to be worth trying with my not-very-mathematical pupils. It's a game for about four to six players or teams and is played on a map made up of a grid of hexagons. The game has two phases — in the first stage the players build competing networks with the intention of connecting towns on the map as efficiently as possible, and in the second phase races are run to make deliveries from one town to another.

Getting Started

First of all we turned our attention to the basic grid. Most children's games are played on a grid of squares, but we found that a grid of hexagons was much better since it allows movement in six directions rather than four (indeed, almost all wargames and adult movement games are played upon such grids). Since I intended to construct the map on a long wall at the back of the classroom we were immediately faced with such questions as "*How many hexagons do we need?*", and "*How big should each one be?*", and soon we had a hexagon factory going full blast. I had been a little nervous about whether it would be possible to make them and pin them up accurately enough to give a satisfactory result, but my fears were groundless, and by the time we'd marked the hills and rivers with appropriate sugar paper we had a very handsome map indeed, especially since one child obligingly produced a supply of nice gold card with which to mark the towns. Visitors appeared from all over the school, and we gleefully asked them to work out the total number of staples we'd used!

In fact I've jumped out of sequence, because there was a stage to the work — before we added the various geographical features — where we used an array of totally blank hexagons. I marked three fairly close hexagons as towns and we started imagining ourselves as railway builders attempting to link the towns as economically as possible using the minimum length of track. All building was done linking the centre of one hexagon to the centre of an adjacent one, and we called the cost of each such section £1z (a "zillion" pounds). For example, the network in the diagram links the three towns for a cost of £7z, but a cheaper network can be constructed.

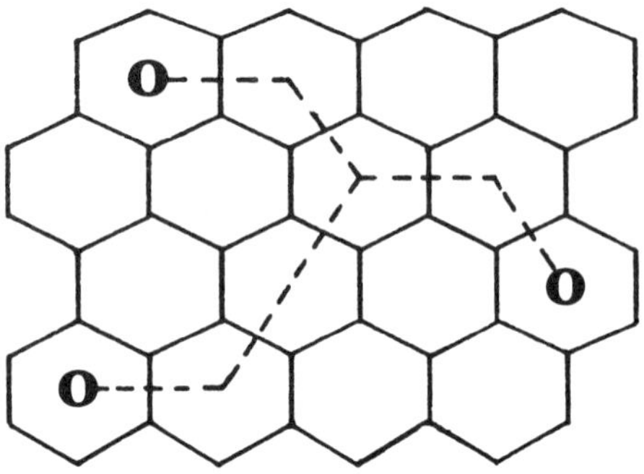

Of course, we looked at the situation if the three towns were in different positions, for maps with four towns, and those with five or even more.

Following this we concerned ourselves with the effects of different geographical features, introducing rivers and hills (rivers always travel along the sides of hexagons, whereas a hill is deemed to fill the entire hexagon — a glance at the map should make this clear). Obviously it is more expensive to build in these areas, so to make a build which crosses a river costs £3z instead of £1z, and it costs £3z to build into a hill and another £3z to build out again.

These few very simple rules have the effect of forcing the building of routes which are very similar to those which were actually constructed by the original companies; wherever possible towns are linked by straight lines, but where necessary quite considerable detours may be made to avoid building in difficult terrain.

There is one other factor which needs to be taken into account. Railway building was not a matter of a benevolent and far-sighted central authority using the best possible route but was an intensely cut-throat exercise with the companies ruthlessly competing to snap up the most economical and profitable routes, so we extended the situation to having two or more companies building at the same time. As a result, the different companies' tracks frequently became entangled, and we introduced a rule to ensure that any company's track must be continuous (in other words, no company could have track which was in disconnected sections), and another so that if any company wished to duplicate a section of track which had already been built by a rival it had to pay £2z for the purpose (though I did allow building *across* other companies' track without extra cost).

We were now ready to embark upon the full simulation. Hills, towns, and rivers were added to the blank map — I particularly wanted to use a reasonably accurate map of our own area (designing such a map is a difficult process, but there are published maps available of many regions of Britain — see Appendix) and thus ensure that not only would we get a decent network of lines but also that the pupils would extend their knowledge of local geography.

Team Work

The class formed themselves into four teams and each team in turn was allowed, starting from London, to build £z worth of track (marked by the stapling of strips of sugar paper in team colours). Over the next few days we had further building rounds (£5z each time) whenever possible — before school, at break and lunchtime, and after school as well; such was the interest that I can't remember any team failing to turn up for a session, however unconventional the time. Gradually the networks took shape, and the spirit of intense competition was encouraged by granting revenues: as soon as a particular town was reached for the first time the successful company received £10z, with later arrivals getting £5z, and this process took place for every town on the map. (This seemed a perfectly reasonable process, for the basic aim of the original builders was to link together towns which needed goods transporting, and although the first arrivals might secure the most business it might still be worthwhile for others to enter even a well-served town. To encourage this still further we allowed companies to build alongside existing track in towns without the usual need to pay for doing so.)

Within a few days track had been built to every town on the map, and the leading company was feeling very pleased with itself. It had become obvious that no company could hope to reach every town, and while one company was strong in one region another might have corresponding strength elsewhere. Now the network was built it was time to test how good the building policies of the companies had been and so we embarked upon the second part of the simulation to see which routes were the best. A giant die was made, and we rolled it twice, using the scores to give the key number (see the map) of one town; then we rolled it twice more to find another town — the numbers were 4, 6 indicating Thame on the map, and 3,2 which corresponded to Beaconsfield. The purpose of these throws was to indicate a contract for making a delivery of goods between these two towns; one company had had the good foresight to build a direct line between the two, so we agreed that they deserved to receive £20z for the contract.

Further contracts were selected in the same way, and in some cases it was possible for two or even three companies to make the delivery. If so, we had simple but exciting races between them — each company made itself a train in its own colour which was placed at the starting town for the contract. Then the companies threw the die in turns, with each moving its train the appropriate number of sections along its track towards the destination town; some deliveries (e.g. from Oxford to Watford, might follow very different routes, but even so the races were normally very close and exciting. Whenever a race was needed we gave the first company to arrive at the destination town £20z, the second £10z, and later arrivals none at all.

We spent a happy half hour or so running trains across the map and noticed that a couple of situations kept cropping up. The first was that frequently a company couldn't make a delivery on its own but could do so if it were allowed to use just a short stretch of another company's track; in this case we thought it perfectly reasonable that they should be allowed to enter the race but that they should pay a hire fee of £1z for each hexagon of track used (and they had to pay this fee whether or not they were successful in the race). The second situation was that it often happened that two companies might both be able to do about half the run but that neither could complete it on its own, and in such cases the two would run a single joint train between them, without charging each other for the use of the track and sharing any profits made.

I forget how many races we held before I called the game to a close, but it was around twenty. One child brought his Polaroid to school to take the official photograph of the directors of the winning company, and several pupils ordered copies of the various readymade maps from the inventor. A year after the event the class — who normally can't remember what they did yesterday — are still recalling it as the best maths topic ever, and some of us are still playing as a group which meets most lunchtimes.

Evaluation

In this account I've really done very little more than scratch

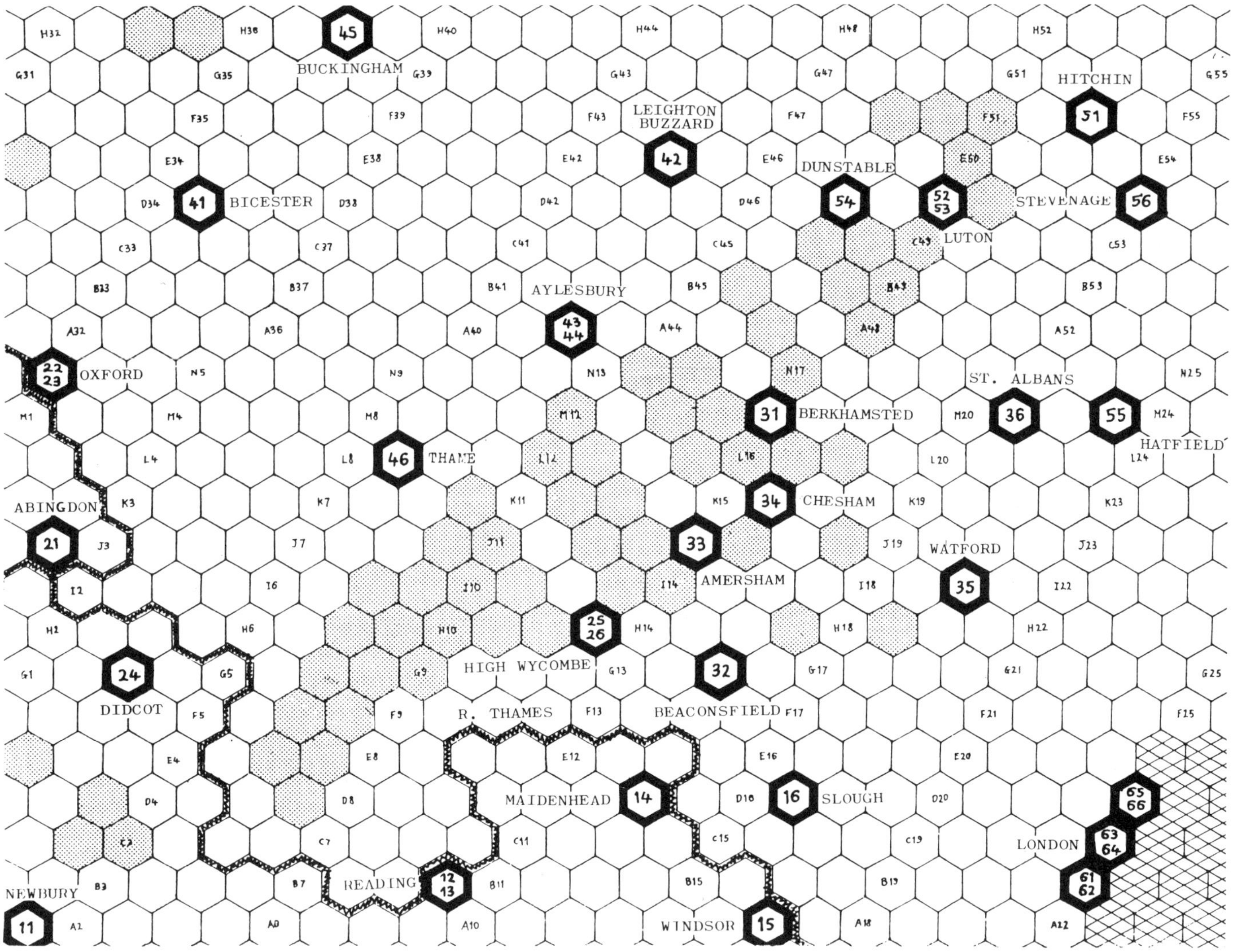

[Code references e.g. F13 are only used in postal play and are irrelevant in face to face play]

the surface of a vast amount of work which was covered, and I could write as much again on a dozen different points — about the experiments we did on several different grids before we settled on hexagons, about work on the 24 hour clock and understanding and creating railway timetables. About the pricing systems for tickets and the tickets themselves, duplicated by the children using a Banda carbon and a fruit jelly.

Or the obvious benefits in local geographical and historical knowledge, highlighted by an unexpected visit from the Divisional Education Officer, who was foolish enough to say that this was all very interesting, but wouldn't an accurate map of what had actually happened be more useful? Fortunately this was exactly what we'd talked about the very day before, and the children were delighted to demonstrate how our map had generated just the same sort of network as had been built in the nineteenth century.

There were other aspects as well. Some teachers would no doubt have wanted to look at railway history, how trains work, the social impact of the railways, and many other topics, but I had quite enough to do without all that.

We kept the economics of the game deliberately simple and even a little woolly, but they worked well enough; and the primitive form of bookkeeping we devised coped — just about — with the problems. Less obvious but equally interesting features stemmed from the way in which the members of each company were forced to work co-operatively and make corporate decisions. This was an aspect I wanted to explore further, and the next year's project (an airline simulation) though nominally an exercise in mastering bearings and journeys brought in short and long term planning, company investment policy, overdrafts, insurance premiums, and more — but that's another story!

Appendix

Railway Rivals was devised by its inventor, David Watts, as an educational activity for pupils of secondary school age; by simplifying and adapting his official rules very slightly I have convinced myself that it has a very great deal to offer children who are considerably younger, and I would like to suggest that the game could provide a focus for a term's work in many junior and middle school classrooms. It is of course not necessary to go to the extent of either making or designing a map for yourself — printed maps of excellent quality (all maps are of A2 size and are covered in plastic so that overhead projector pens can be used to mark the tracks) are readily available from the inventor at low cost. You can buy a map, rules, counters and dice, and a set of pens for less than £5; alternatively you can have a map and a set of rules — and this is all you really need — for less than £2. Indeed, many of the maps are also available as uncovered paper for under £1, which makes them a remarkable bargain.

About a dozen different scenarios are available in this ready-printed form, including regions of England, Scotland, and Wales. There is also an Irish map, a French one, and at least three different areas of the United States. Moreover, there are many more unofficial maps ranging from the North Pennines and East Anglia to Spain, Romania, and even Tolkein's Middle Earth, so it is quite likely that the inventor will be able to help you obtain a set of instructions for making up a map for which a ready-printed one isn't available.

The inventor is always happy to receive enquiries, by post or telephone, and will supply a catalogue of Railway Rivals items and of his many other games on request: David Watts, Rostherne, 102 Priory Road, Milford Haven, Dyfed SA73 2ED.

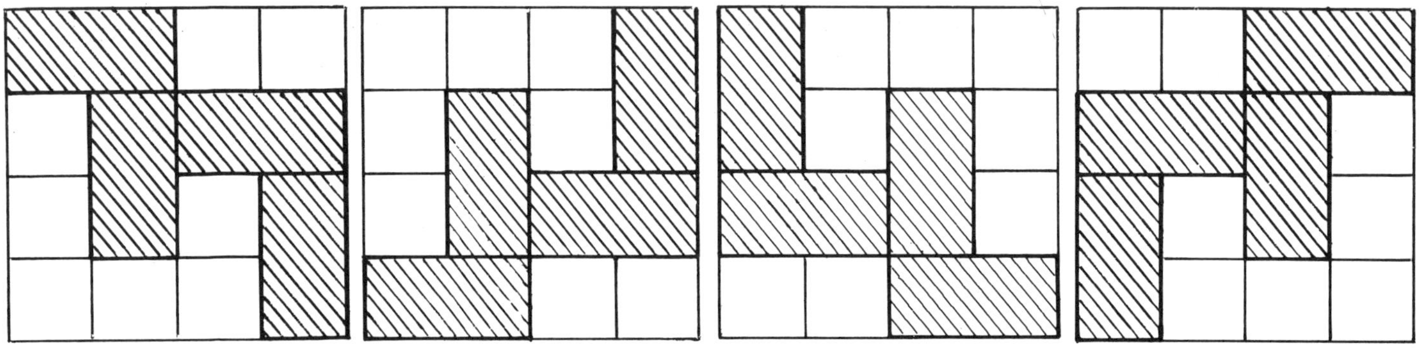

Placing four tiles 2×1 on a 4×4 square: an investigation

by Beryl Boyd, Primary School Consultant, ILEA

Starting Point

How many different ways can you place four tiles 2×1 on a 4×4 square? Assumed rules: Tiles must not overlap, and each tile must occupy two whole squares vertically or horizontally.

e.g. not

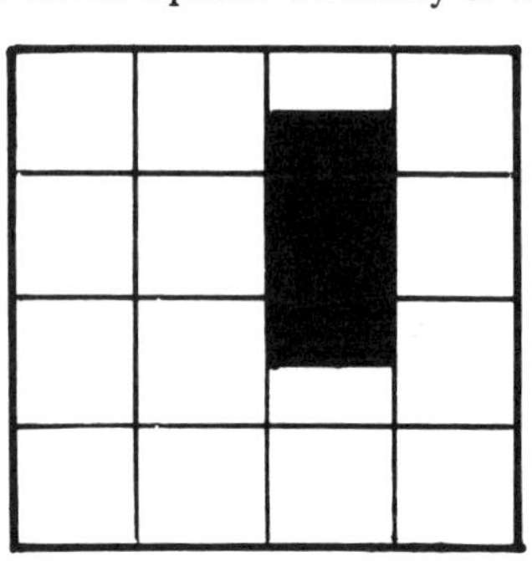

I started marking out 4×4 squares on squared paper and colouring possible positions of tiles. When I had made about 40 patterns, I stopped to review the results.

Some were rotations or reflections of others; should I count these as different?

Because I had coloured the positions of the tiles, the direction in which they had been placed was ambiguous in some cases.

e.g.

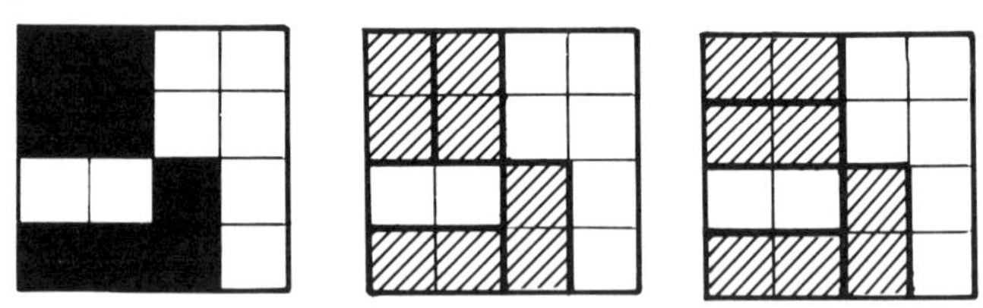

A B

Had the square been made by A or by B?

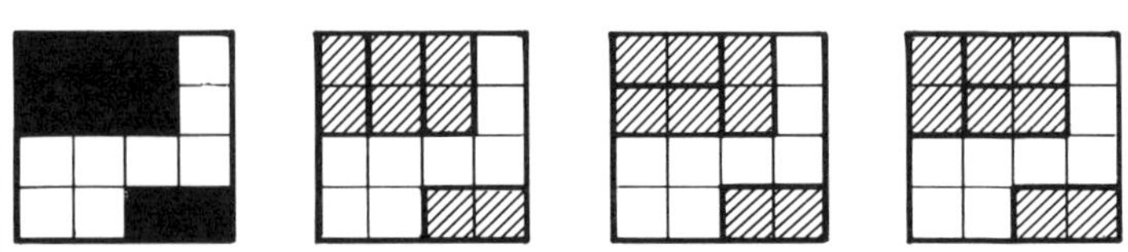

A B C

Had the rectangle been made by A or by B or by C?

In other words, each time I made a square from two tiles or a rectangle from three tiles, I needed to make it clear in which direction I had placed them and also to include any possible variations. From this stage onwards I decided to shade rather than colour.

Groups of patterns were emerging;

e.g. those including a 2×2 square
those including a 3×2 rectangle
those including a 4×1 rectangle, etc.

How many different patterns were there going to be?

Would I be able to tell when I had found them all? I decided to make a chart, numbering the possible positions of the rectangular tiles

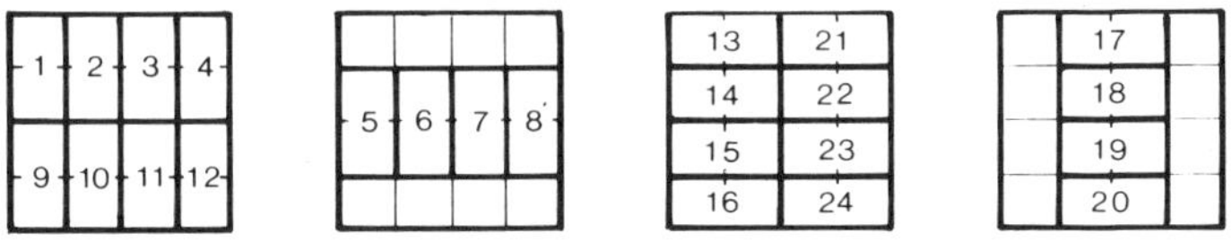

I started to list possible number combinations:

(1 2 3 4),
(1 2 3 5) — not possible because 5 would overlap 1.
(1 2 3 6) — not possible because 6 would overlap 2.
(1 2 3 7) — not possible because 7 would overlap 3.
(1 2 3 8)

I stopped, and made a list of impossible combinations

1 would overlap 5, 13, 14
2 would overlap 6, 13, 14, 17, 18
3 would overlap 7, 17, 18, 21, 22
4 would overlap 8, 21, 22
5 would overlap 9, 14, 15
6 would overlap 10, 14, 15, 18, 19
7 would overlap 11, 18, 19, 22, 23
8 would overlap 12, 23, 24
9 would overlap 15, 16 and so on.

I resumed work on my list of *possible* number combinations, omitting the impossible ones or crossing them out if I had included them in error.

(1 2 3 4), (1 2 3 8), (1 2 3 9), (1 2 3 10), (1 2 3 11), (1 2 3 12) . . .

Several days and two exercise books later, I reached (21 22 23 24).

I started to draw the patterns beginning with the most obvious outline, a rectangle 4×2.

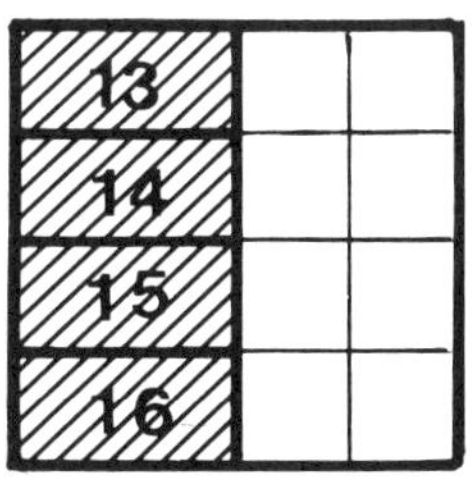

and reflection

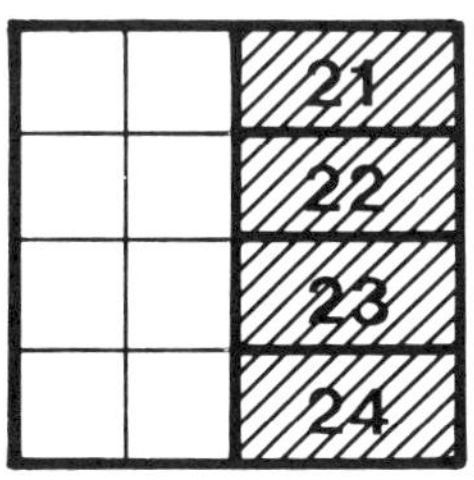

(13, 14, 15, 16) (21, 22, 23, 24)

by rotating them ¼ turn

(1, 2, 3, 4) (9, 10, 11, 12)

I crossed out these sequences from my list, noting that by rotating the patterns a ½ turn the sequences would re-occur.

I investigated different ways of making the same rectangle and crossed off the combinations representing the identity, rotation and reflection of these.

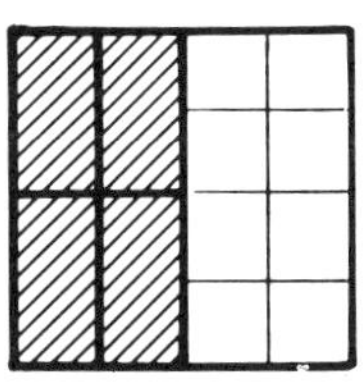

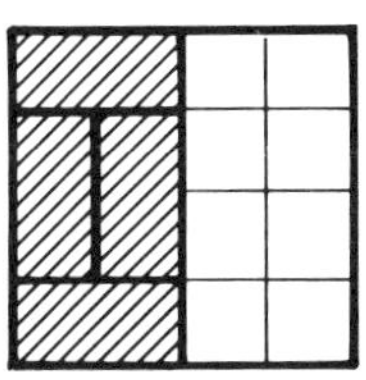

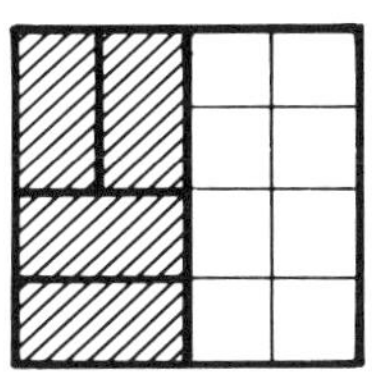

Then I made the position of the rectangle central along one side; this led to ways of representing;

2 squares 2×2
2 rectangles 4×1
1 rectangle 3×2 and 1 rectangle 2×1
1 square 2×2 and 2 rectangles 2×1
1 rectangle 4×1 and 2 rectangles 2×1

All these I drew with their reflections, and crossed off the number combinations which they and their rotations represented.

At this stage I was left with many more unrepresented number combinations than I had anticipated, and I had a file full of patterns.

Part II of the Investigation

Ways of Placing Four Rectangles 2×1 on a Square 4×4, Without Making a Square or Larger Rectangle

As I came to patterns which were the same if rotated, I collected them together and worked through the rest of my number combinations. (The supply of squared paper was a recurring frustration.)

At the end I had drawn more than 600 patterns (many were reflections) and with their rotations they represented more than 2 500 number combinations. I kept a record of the numbers representing each pattern, in a book with pages numbered the same as the file of patterns.

I found the "perfect" pattern in;

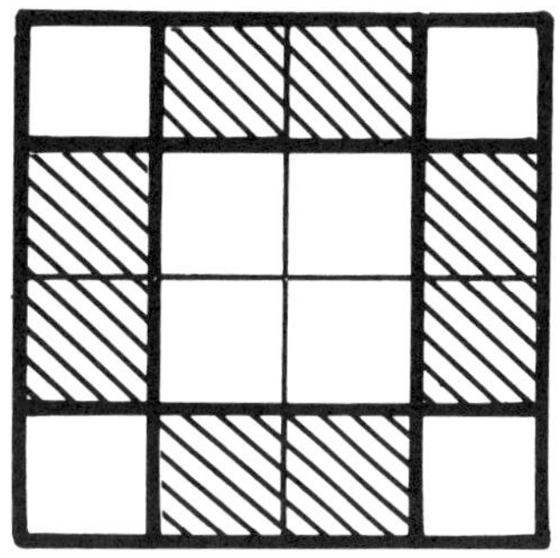

(5, 8, 17, 20)

Rotation = reflection = identity.
A pattern with vertical, horizontal and diagonal symmetry.

Activities for Children

(I have tried these in primary classes)

Draw a square 4×4 (e.g. on 2 cm squared paper).

Cut out four rectangles 2×1 from a separate piece of paper. Colour your rectangles one colour.

1. Place the rectangles within the square so that each rectangle covers two whole squares, and the rectangles do not overlap.
   Record the position of your rectangles by shading part of the square.
   Repeat several times, changing the position of the rectangles each time.
   What fraction of the 4×4 square have you shaded each time?
   What area have you shaded each time?
2. Work with a partner.
   One person places three of the rectangles according to previous rules.
   In how many different positions is it possible for the second person to place the fourth rectangle?
   How can the first person make it easier/more difficult for the second person to find several ways?
3. Work with a partner — two rectangles each.
   First person places two rectangles within square.
   Second person must place his/her rectangles so that the total shape is symmetrical in some way.

Where to go next?

What would happen using a different size square
and/or different size rectangles
and/or different number of rectangles.

What would happen using rhombi within an equilateral triangle?

What would happen if. . . . . ?

MATHEMATICS COMPETITIONS

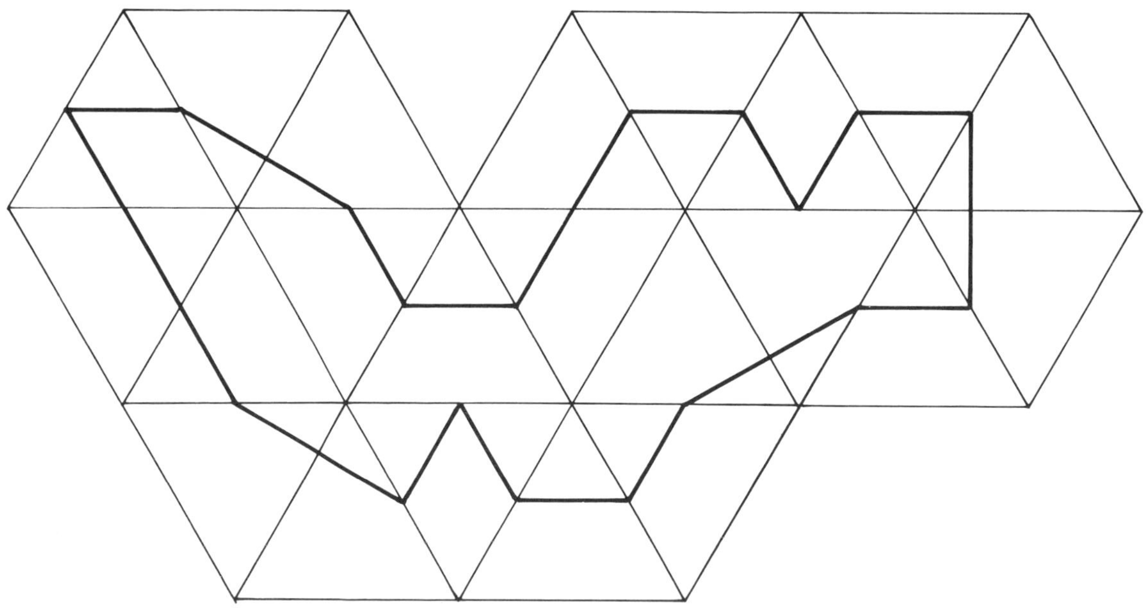

an alternative

by Keith Porteous
Newland High School, Hull

Mathematics competitions for school-children are typically of a type that may be referred to as "excellence-orientated". That is, they consist of a set of problems, possibly graded in terms of difficulty, which are designed to test mathematical ability in novel and interesting ways. The problems are generally of an all-or-nothing character — you either do what is asked or you fail to do what is asked. Further, the problems are demanding: some may be harder than others, but all require some thought, none are straightforward. One can be sure that the successful competitors are the most able mathematicians, and thus it is that the pupils who are attracted to such competitions are those who are capable, and furthermore confident enough to put their abilities to the test. The clientele is not large. The annual paper set by our own Mathematical Association, the winners of which take part in the international Mathematics Olympiad, falls into this category, as does the Liverpool-based Challenge[1] and the Australian Mathematics Competition[2], both of which have been reported recently in Mathematics In School.

But is it not possible to design a competition which will appeal to a majority of pupils — one which is enjoyment-orientated, where all the competitors feel that they have succeeded? Such a contest need not produce the best mathematician as winner, but as an exercise in fostering good feelings for the subject it may well be worthwhile.

At Newland High School a termly competition has been held along these lines for the last six years. The aim is simple: to attract as many entrants as possible. Thus competitions are constructed with the following principles in mind (though it has to be admitted that not all are fulfilled each time):

1. The task set will be a single project, perhaps with related stages, not a series of separate problems.
2. The task should be open-ended as far as possible, so that the idea of finishing the task, or solving the problem, is fairly meaningless.
3. There should be room for decision-making, or creativity, on the part of competitors.
4. The project must be within the capabilities of all pupils in the school, from the least able 13-year olds to the brightest 16-year olds. (Sixth-formers are excluded — they have other jobs to do, as will be seen later.)

Three broad types of competition have emerged over the years.

The first emphasises the creative element. In one, entrants were to design an Easter-egg box, or other interesting three-dimensional object, with its net. In another, they were to construct a cross-number. The most successful in this category was run in conjunction with the Art department: entrants were to draw or paint an Escher-type picture of a tesselating shape. About 80 pupils submitted entries for this.

The second category consists of the end-of-the-summer-term tournaments. Each year a new game is used, and the tournament is based on a mini-league system (like the World Cup) so that all entrants will play several games before the knock-out stages are reached. The games used have varied considerably in nature: some are adaptations of well-known games, some are home-grown. One of the latter is called "Chains". Each player has a copy of the chart shown.

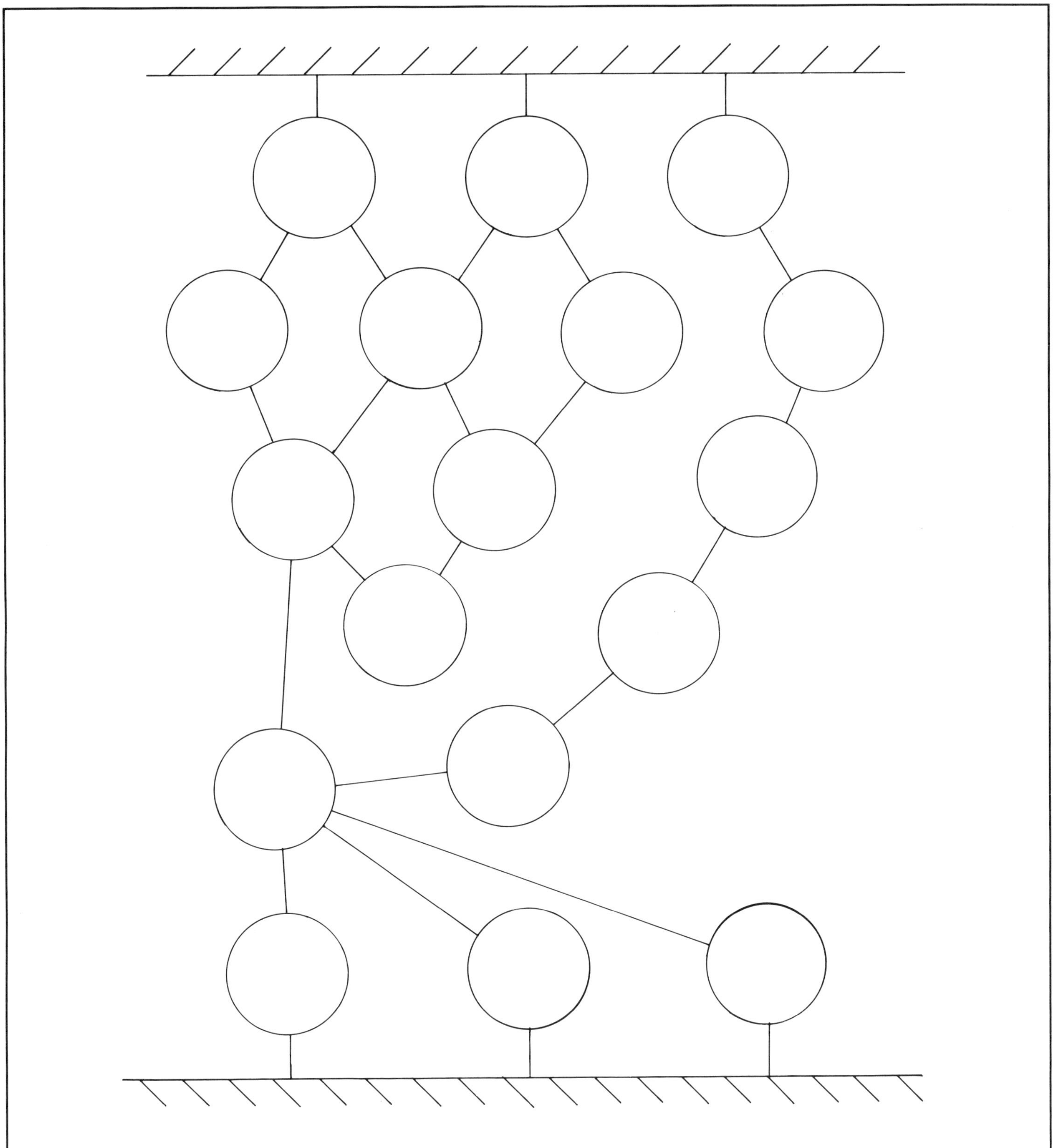

Fig. 1

The pupil's first task is to write the numbers from 1 to 17 in the cells in any way she chooses. Play then proceeds by taking turns throwing three dice: the numbers on the dice are to be combined, using normal mathematical operations, to produce a result in the range 1–17, that number then being crossed off on the player's chart. Only one number can be deleted each turn.

Thus, if the dice show the numbers 2, 3, and 6, the player may delete 11 $(=2+3+6)$ or $1(=6-2-3)$ or $6(=3+\frac{6}{2})$ etc. The winner is the first player to delete a complete chain of cells linking the top of the chart to the bottom. As skill with this game develops the strategic positioning of numbers in the cells becomes important. The game can be made more difficult by using other numbers, say from 18 to 34. What about the first 17 prime numbers?

All the competitions mentioned so far have been held outside of normal lesson time. In recent years, though, an in-school competition has been held at the end of the Autumn term. (This is the only occasion on which the Mathematics department claims lesson time from the rest of the school!) Each of these has centred around a particular topic — a developing set of exercises dealing with hexiamonds, or with pentominos; fitting together equilateral triangles to make a polygon of 3 sides, 4 sides, 5 sides, and so on (how far can you go? how many triangles do you need for an n-sided polygon?). It is perhaps worth describing one of these projects in detail.

Competitors were given a card, the bottom part of which

Fig. 2

was a score sheet which was updated by the supervisor as progress was made. The top part bore 24 triangles as shown below — these were first to be cut out.

The task is to join together some of these triangles (edge to edge and vertex to vertex) in such a way that the heavy lines join to form a closed triangle; then a quadrilateral; then a pentagon, and so on as far as possible.

For example:

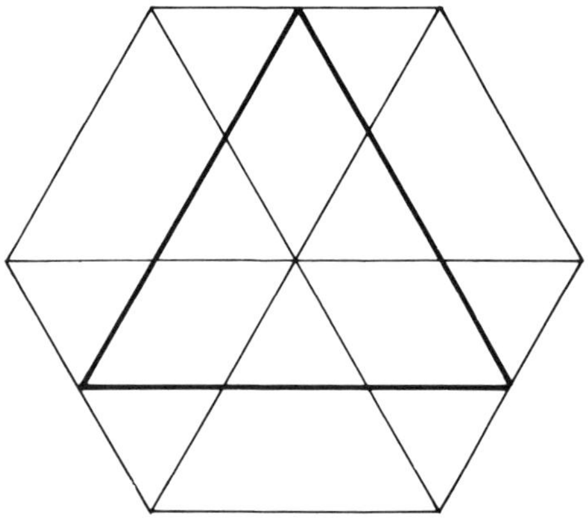

...this is a triangle ...

...and this 18 gon is shown in the heading

So, the aim is to construct polygons of an increasing number of sides. Up to 19? 20? 21?

To make the administration of these in-school competitions easier sixth-form A-level Maths pupils are drafted in to act as supervisors. This is found to be necessary, since these competitions attract up to 400 entrants (mostly working in pairs), which is more than half the school — very pleasing, for a girls' comprehensive.

Excellence-orientated competitions are fine, but if enjoyment-oriented events can involve so many people with the subject in a positive and rewarding way then they, too, have a part to play.

References

1. Porteous, I. & Veevers, A. (1982); Challenge on Merseyside. Mathematics in School, 11.2.
2. Edwards, J. (1985); Boys and girls in the Australian mathematics competition. Mathematics in School, 14.5.

MATHS GAMES

WORKSHOP

Dave Kirkby *Part Four* Sheffield Polytechnic

Take a board . . .

3	8	1	6	1
2	4	8	2	5
7	0	5	3	9
2	7	4	9	3
1	4	6	7	0

4	9	5	6	2
7	1	8	9	6
4	0	5	0	3
7	5	1	3	7
9	2	8	6	8

Make two game boards by drawing square 5×5 grids. Write random arrangements of digits in the boxes.

These boards can then be used for a variety of number games. Here are some examples. Readers are invited to suggest other possibilities.

All the games are for two players both of whom have a set of counters of their own colour. Games 1 to 3 require the use of one board only.

Game 1

Use two dice, each numbered 1 to 6.
Players alternately throw the two dice and find the sum. A counter can then be placed on any square containing that sum.
If the score is 10, a counter can be placed on 0.
If the score is 11, a counter can be placed on 1 and, if the score is 12, the player can have another turn.

(a) The first player to obtain a row of 4 counters (horizontally, vertically or diagonally) wins, or
(b) Play 20 rounds each. At the end of twenty rounds sum the numbers in the uncovered squares. The player with the lowest total wins.

Game 2

Use two sets of cards, each numbered 1 to 10.
The cards are shuffled and placed face down in a pile.
Players agree on a two-digit target number, say 42.
Player A turns over the top card, suppose it is 8. He divides 42 by 8, finds the remainder, i.e. 2 and places a counter on a square containing 2.
Player B then takes his turn. If a counter cannot be placed — tough luck! The game finishes when all 20 cards have been revealed.
(a) The player who has placed the most counters is the winner, *or*
(b) Players add up the numbers covered by their counters, and the one with the largest total wins.

Game 3

Decide on a multiplication table to be practised, say "4 × table".
Player A places a counter on any square — suppose it covers up the number 5. Then he says, "Five fours are 20" and records 20 as his score.
Player B then places a counter adjacent to the last one played — suppose it covers the number 3. He says, "Three fours are twelve" and records 12 as his score.
Players continue alternately placing counters adjacent to the last one played and recording the score. When the board has been filled, the winner is the player with the largest total score. If a player is unable to place a counter on an adjacent square, he can place his counter anywhere available on the board.

Games 4 to 9 require the use of two boards, one for each player.

Game 4

Use two dice, each numbered 1 to 6.
Players alternately throw two dice and find the sum. Suppose it is 8. A counter can then be placed on a square containing the number 8, *or* a set of squares whose numbers sum to 8, e.g. 3, 4 and 1.
If a player cannot do this exactly then he must do nothing at all on that turn.
The winner is the first player to cover the whole board.

Game 5

Use two dice, each numbered 1 to 6.
Each player in turn throws the two dice. A counter is placed on a square containing either the SUM or the DIFFERENCE between the two scores on the dice.
The first player to cover his whole board wins.

Game 6

Use two sets of cards, each numbered 0 to 9.
The sets of cards are shuffled and placed face down in two piles.
Player A reveals the top two cards, e.g. 7 and 5, and multiplies the two scores together, i.e. $7 \times 5 = 35$.
His score is the digit in the units' place 5 so he places a counter on a square numbered 5.
Then Player B takes his turn, revealing the next two cards. Play continues this way and, if the piles of cards are used up, they are reshuffled and replaced in two piles.
The winner is the first player to complete:
(a)the board, *or*
(b)a row of four (horizontally, vertically or diagonally).

In each of games 7, 8 and 9 two dice are required (one representing the tens digit, the other the units digit of a two-digit number).

Game 7

Players take turns to throw the two dice to reveal a two-digit number. A counter can be placed on a square containing a number which is a DIVISOR of the two-digit number.
For example, suppose the dice reveal 32. Then a counter can be placed on a square containing a 1, 2, 4 or 8. Each player can only place one counter on each turn.
The winner is the first player to cover
(a)the whole board, *or*
(b)a line of four (horizontally, vertically or diagonally).

Game 8

Each player places counters on the squares of his board which contain a zero.
Player A places a counter on any square, say 3.
He then throws the two dice to reveal a two-digit number, say 32. He scores the remainder when 32 is divided by 3, i.e. 2. The score is recorded.
Player B then has his turn to choose a starting number, throwing the two dice and scoring a remainder.
The game finishes when each player has had ten rounds.
The winner is the player with the largest total score.

Game 9

Players take turns to throw the two dice to reveal a two-digit number. Suppose it is 32.
Counters can then be placed on squares containing factors of 32, i.e. 4 and 8, since $4 \times 8 = 32$.
Alternatively, counters can be placed on 2, 4 and 4, since $2 \times 4 \times 4 = 32$.
If a player cannot do this *exactly* then he must do nothing at all in that turn. Play ten rounds each. The winner is the player who has placed the most counters.
Variation: Change the numbers on the dice.

Coloured Squares

One variation to this collection of games is to colour some of the squares on the board.

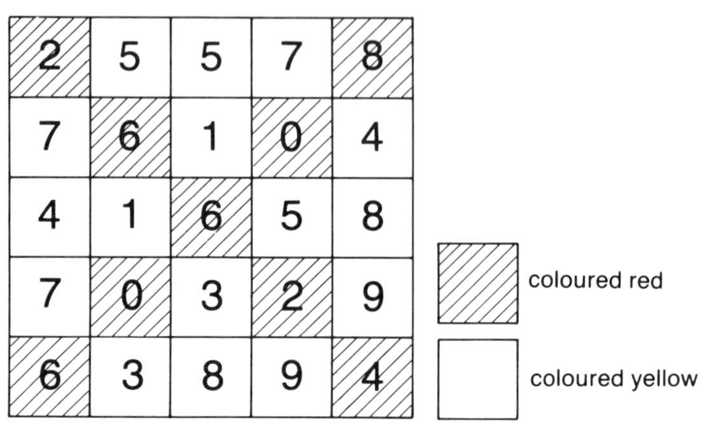

This allows the possibility of scoring systems based on the colour of the square. (As, for example, the double word scores, triple letter scores in "SCRABBLE").

References

Kirkby, D., (1985) *The Eigen Square*. Eigen Publications.

A GRID GAME

by J. R. Branfield

In this article, Mr Branfield shows how a simple game, suitable for top Juniors, can be developed to sixth-form level and, at the same time, provide a creative investigation with varied mathematics.

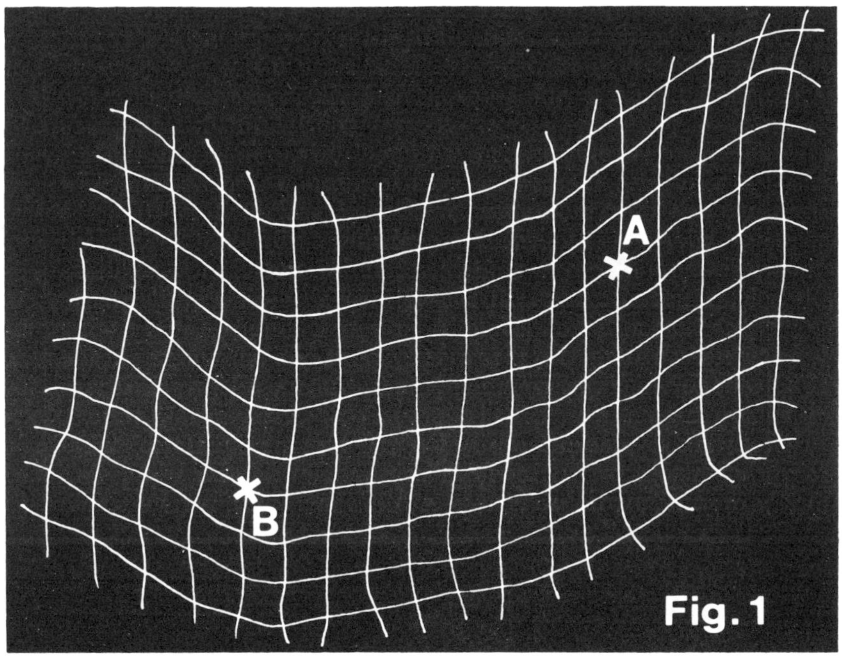

Fig. 1

Gridville is a town whose roads have all been constructed to meet at crossroads and each street from crossroad to crossroad is the same length. Houses occupy all the spaces. I live on a crossroad at *A* and my friend lives at *B*; Fig. 1.

We play a game together. I decide on a rule of moving among the streets and I must use that rule whenever my friend has moved. A helicopter keeps us informed of our movements and the game is for my friend to discover what rule of movement I am using and then to move so that we reach the same place at the same time. We cannot go through the houses in the spaces — only along the streets and we always move a whole number of streets (not fractions of a street).

As a first rule I choose *"twice as far in the opposite direction."* My friend starts and moves one street westerly from *B,* so I move two streets easterly from *A,* and after another move my rule is soon discovered. To meet at *C* our travels are: (Fig. 2).

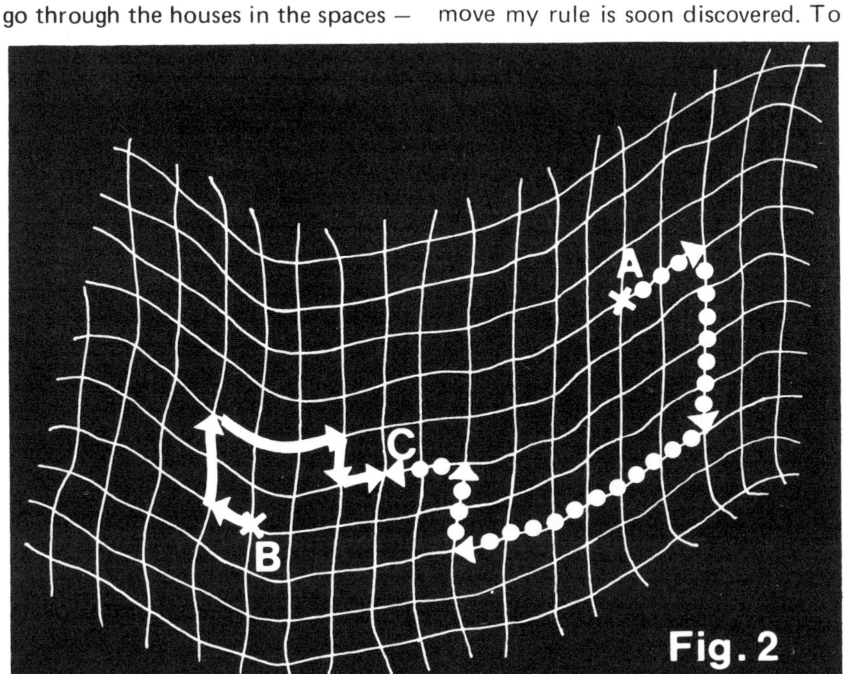

Fig. 2

I travelled 16 streets, how many did my friend travel? Could I have reached *C* using fewer streets?

We meet at *C* often under this rule – what about using the same rule but this time we are to change our travels and meet at *D* (Fig. 3).

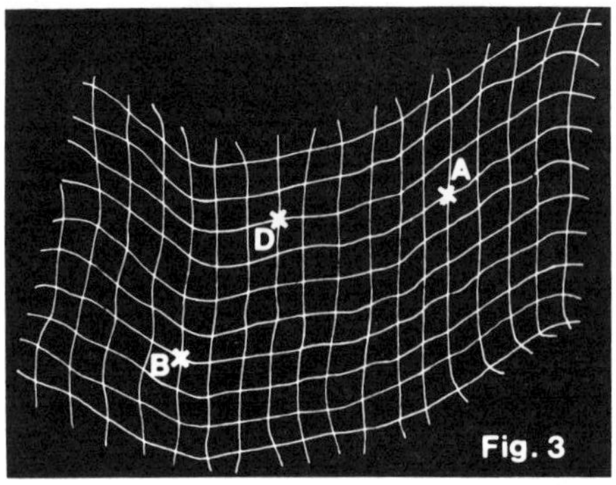

Fig. 3

After several attempts and chess-like thinking of "if I go there, he will be there" it becomes apparent that we **cannot** meet at D. *C* is the only possible meeting point for this rule and is thus an invariant point. That it is invariant can be argued from the rule; "in the opposite" means we must meet **within** the "rectangle" formed by *A* and *B* otherwise we move apart, and "twice as far" gives the proportions for the meeting point.

What about the rule *"twice as far in the same direction"?*

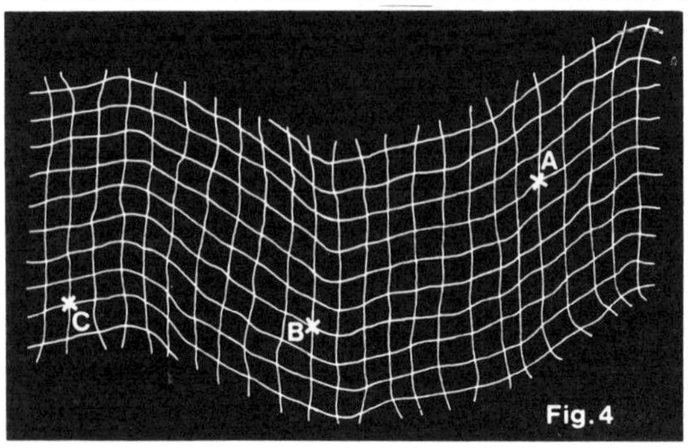

Fig. 4

Again *C* is the only meeting point (and not any of the three other corners); Fig. 4.

I have changed the rule of moving; what is it from. Fig. 5? (*B* moved first).

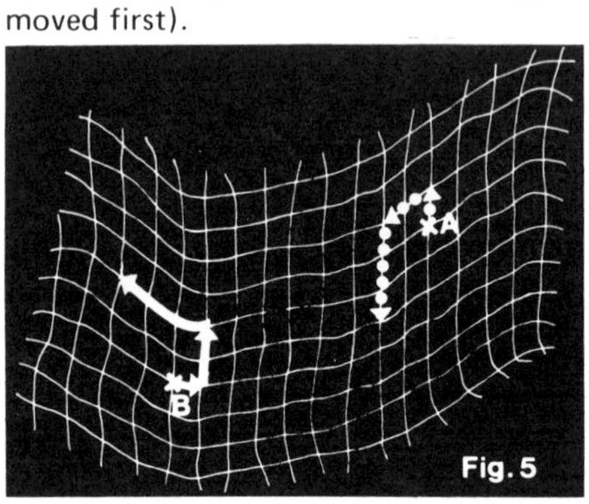

Fig. 5

"Through the same distance, 90° anticlockwise." Has this rule an invariant meeting point?

There are obviously a large variety of rules which can be invented and plenty of games to play.

Now make the grid into a co-ordinate square lattice with *B* the origin of axes (not necessary for the mathematics but familiar to see and refer to!) Given the co-ordinates of *A* and the rule, can the co-ordinates of the meeting place be found? Using the first rule and with coordinates of *A* as (9, 3), the movements of *B* and *A* can be determined by a set of linear equations,

$x' = 9 - 2x$

$y' = 3 - 2y$

where (x, y) is the position of *B* and (x', y') the corresponding position of *A*. *B*'s first move was to $(-1,0)$, so the position of *A* now is

$x' = 9 + 2 = 11$

$y' = 3 - 0 = 3$

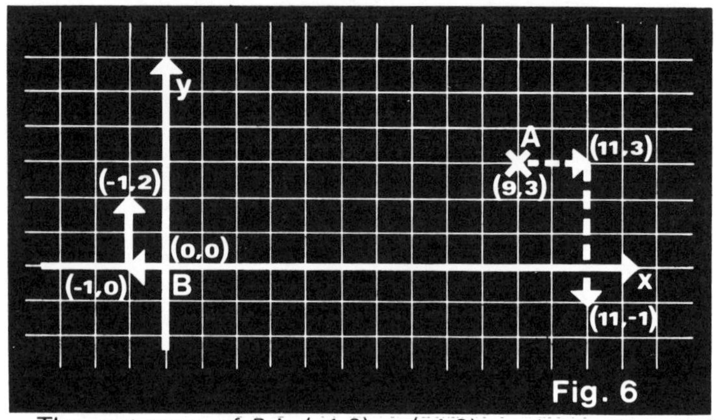

Fig. 6

The next move of *B* is $(-1,0) \to (-1,2)$, so the position of *A* would be $(9 + 2, 3 - 4)$; i.e. *A* moves $(11,3) \to (11,-1)$.

The meeting point is given when the co-ordinates of (x,y) and (x',y') coincide, giving the equations

$x = 9 - 2x$

$y = 3 - 2y$

to solve uniquely for *C* as (3,1).

For the rule, *"through the same distance, 90° anti-clockwise"* the equations which describe the movements are

$x' = 9 - y$

$y' = 3 + x$

giving the invariant meeting point as (3,6). When the rule may be determined by a set of linear equations it is possible to find the meeting point from the solution of the equations. (The initial positions of *A* and *B* as well as the rule govern the position of the meeting point – it may be part way along a street or (inaccessible) in a space). It is also possible to find *equations* which have **no** meeting point, e.g.

$x' = 4 - y$

$y' = 3 - x$

or those which have many, e.g.

$x' = 3 - y$

$y' = 3 - x$

and to interpret the rule in statement form in the case of these equations and see how it acts in Gridville.

A set of linear equations may be represented by a matrix. It is therefore possible to interpret the first rule as

$$\begin{pmatrix} x' \\ y' \end{pmatrix} = \begin{pmatrix} -2 & 0 & 9 \\ 0 & -2 & 3 \end{pmatrix} \begin{pmatrix} x \\ y \\ 1 \end{pmatrix}$$

$$= \begin{pmatrix} -2 & 0 \\ 0 & -2 \end{pmatrix} \begin{pmatrix} x \\ y \end{pmatrix} + \begin{pmatrix} 9 \\ 3 \end{pmatrix}$$

$$= 2\begin{pmatrix} -1 & 0 \\ 0 & -1 \end{pmatrix} \begin{pmatrix} x \\ y \end{pmatrix} + \begin{pmatrix} 9 \\ 3 \end{pmatrix}.$$

The factor 2 is an enlargement value and the matrix $\begin{pmatrix} -1 & 0 \\ 0 & -1 \end{pmatrix}$ represents a rotation through 180° about the origin. These geometrical properties are shown in the paths of *A* and *B*. The rules giving linear equations may be interpreted in terms of translation, reflection, rotation and enlargement transformations.

Writing the second rule in matrix form as

$$\begin{pmatrix} 0 & -1 & 9 \\ 1 & 0 & 3 \\ 0 & 0 & 1 \end{pmatrix} \begin{pmatrix} x \\ y \\ 1 \end{pmatrix} = \begin{pmatrix} x' \\ y' \\ 1 \end{pmatrix}$$

we can see that finding the meeting point is seeking the eigenvector of the matrix (unique from the unit entry) with corresponding eigenvalue 1.

The general analysis of

$$\begin{pmatrix} a & b & h \\ c & d & k \\ 0 & 0 & 1 \end{pmatrix} \begin{pmatrix} x \\ y \\ 1 \end{pmatrix} = \begin{pmatrix} x \\ y \\ 1 \end{pmatrix}$$

establishes that an eigenvalue of 1 exists and that providing $ad - bc \neq a + d - 1$ there will be a meeting point.

(For a different approach to this situation see *Analytical Methods in Transformation Geometry* by G. Giles in *Mathematical Reflections*).

7 Research, Political and Educational Themes

by Peter Reynolds

The 1970s witnessed a variety of political and educational issues, often aired very publicly. Perhaps the most significant issue was the alleged inadequacies of school leavers highlighted in the great debate initiated by the Prime Minister, James Callaghan, in his famous 1976 Ruskin speech, which led directly to the formation of the Cockcroft committee of inquiry in 1978. The debate had been anticipated in *Mathemetics in School*: the Institute of Mathematics and its Applications (IMA) had through its conferences, mischievously titled 'Mathematical Shortcomings at the School/Employment Interface', contributed to the undermining of public confidence. The Cockcroft committee was later to find most of these criticisms were either untrue or difficult to substantiate: the flavour of the times was well captured in the article of D. R. Green in January 1976 and picked up in an Editorial in the next issue.

There were many general themes which made repeated headline appearances during the mid 1970s; these included falling standards, difficulties in the supply of teachers, 'modern mathematics', mixed ability teaching, complaints from industry about the inadequacies of school leavers and a growing concern about the dangers to the quality of pupils' thinking posed by the new technology of calculators and computers. There were lots of changes in the air which were nicely pulled together by Michael Cornelius in March 1975.

The main attempt to grapple with 'standards' was the setting up of the Assessment of Performance Unit (APU) which started a programme of testing 11 and 15 year old pupils in 1978; the Editorial in the January 1978 issue reserved judgement on its value. The intention behind the APU was that the government would be able to measure improvement in standards over a period of time. Ironically, the main use of the APU work, together with the other major research project in the 1970s to look at performance — Concepts in Secondary Mathematics and Science (CSMS) — was to enable the Cockcroft committee to identify the now famous 'seven year difference' at age 11 and to recommend a differentiated curriculum at 16.

Although the CSMS report *Children's Understanding of Mathematics: 11–16* was not finally published by John Murray until 1981, readers of *Mathematics in School* were treated to some delightful and powerful articles well before the final report; we have chosen *Is it an 'add' Miss?* by Margaret Brown and Dietmar Küchemann from November 1976 as representative. Here, at last, was careful research evidence on what children can do — or, more importantly, appear not to be able to do.

There were several curriculum development in the 1970s — both primary and secondary — although the emphasis was not on the *modern mathematics* of the 1960s. The influential and, later, slightly controversial *Fletcher* primary series, with the title *Mathematics for Schools*, was launched in 1971 and was reviewed by Hilary Shuard in November 1972 long before cheap calculators which now affect her thoughts were available. In fact, generally, the mid-1970s saw a vigorous reaction against new ideas, mainly secondary, although *Fletcher* maths was often included. SMP, which had flourished since its launch in 1961, became the subject of strong criticism (for example) by D. M. Burley in November 1975 and was stoutly defended (for example) by the newly appointed Executive Director of SMP, John Hersee.

Finally, and topically, we end this rather fragmented selection with an article by Joe Watson from September 1978 which anticipated the problems of testing associated with the national curriculum. He provides a timely reminder that we should 'not be deterred from trying to give children a wider view of mathematics than the arid *basic core*'. Let us hope that teachers will have the confidence to provde mathematically rich courses whatever tests are imposed; the chances are good that, given a stimulating course, children will do better on any test.

NUMERACY putting the clock back?

by D. R. Green C.A.M.E.T., Loughborough University

Some thoughts on the second conference on "Mathematical Shortcomings at the School/Employment Interface", held at the Shell Centre, Nottingham University, 8—10 July 1975, convened by the Institute of Mathematics and its Applications.

This conference (which I shall refer to as "Interface II") was a follow-up to that held just one year earlier (July 1974). A very full report of the first conference (referred to as "Interface I") has been published by the I.M.A. (Ref. 1), and a personal account of its proceedings was included in *Mathematics in School*, Vol. 4 No. 1, (Ref.2).

The 1974 conference grew out of repeated criticisms of school mathematics by those in industry concerned about the standard of numeracy of their recruits. Interface II cannot be properly appreciated without some reference to Interface I and so I make no apology for spending some time in commenting on the earlier conference. I believe that Interface I came as something of a shock to many, like myself, concerned with secondary mathematics. The industrial representatives (mainly from the Engineering Industry) made out a good case for more serious consideration of the arithmetical skills of those pupils, probably studying for CSE in mathematics, who are to become Engineering trainees—and many thousands of them are, too! (23,000 in 1974). And this was not to imply that many other areas of employment do not require basic mathematical skills (e.g. nursing), but they were not well represented at that conference. "Modern mathematics", both content and method, was blamed by those from industry for a recent fall in the standard of their entrants. There followed a disturbing but perhaps necessary verbal battle between the "industrialists" and the "school teachers". Professor J. V. Armitage performed a small miracle in controlling the debates and extracting a measure of agreement, and indeed a Core Syllabus was agreed upon (Ref. 3).

The "shock" which I received at Interface I, and to which I alluded above, stemmed from the realization:
(a) that school mathematics syllabuses were not, after all, so entirely divorced from the real life of many of those leaving to enter employment at age sixteen.
(b) that I had no idea what mathematics *had* proved useful to those CSE pupils whom I had taught in previous years, and that I had not been drawn to think of this seriously before.
(c) that the industralists looked upon schools as places of "training" and were not satisfied with the teachers' ideas of "general education".
(d) that the teachers themselves were apparently hiding in their ignorance behind generalizations and unconvincing theory. (I for one felt not a little chastened for so long having accepted the simplistic view that "modern mathematics" is superior to "traditional mathematics".)
(e) that those from industry were unsympathetic and ignorant of the problems of school mathematics and wished to treat Education as just another industry amenable to the usual economic analysis.

All parties at Interface I agreed that another conference was needed the following year, and so we found ourselves assembled at Interface II. It is useful to look again at the reason for meeting. Indeed, what *is* the problem?

A recent edition of *Skill*, a paper published by the Engineering Industry Training Board (Ref. 4), can be quoted from as showing how those in one part of industry see it:-

(i) "... all kinds of firms say they find that school-leavers can't do the arithmetic which trainees (a) need for training, (b) used to take for granted. The necessary 'remedial maths.', says industry, is an unnecessary extra cost of training time and money."

(ii) "The educational world simply does not understand why it is necessary to be capable of handling (both) imperial and metric numbers freely."

(iii) "... What the customer wants matters to a firm much more than somebody's unauthorized metrication policy".

(iv) "Employer interviewing: "How many 64ths in an inch?' Traineeship candidate: 'I don't really know, Sir, but there must be quite a lot'."

(v) "a working party ... report highlighted the usual weaknesses ... Decimals. Fractions. Division of whole numbers. Percentages. Ratios. Transposition of formulae."

(vi) A previous edition of *Skill* (Ref. 5) included a letter from a company manager who wrote of the steady decline in school-leaver performance in recruitment tests for apprentices:

"... This year I collected all the scraps of paper bearing their rough working, in an attempt to follow the devious workings of their little minds. It was revealing and utterly scandalous to see the depths of their ignorance in simple fraction and decimal matters. Many fail to place the decimal point in vertical line when adding or subtracting. Others do not know their basic tables and evaluate 7×9 by writing down 9 sevens in a vertical column then adding them up—frequently getting it wrong in the process. Perhaps it would be kinder to say "untutored" rather than "tiny" minds for much of the fault must lie in the teaching.

Question 23 produced ·000327, 3·22, 3·142, 2·17 as π to 4 places of decimals. God help us, those came from boys who reached our short list!!! The "rejects" came up with 00.22, 3.1440, 2.2111, .33, $\frac{22}{7}$, 1·432, 1 342 etc. etc. *Question 26* produced 29, 37, 40, 45, 50, and 80 inches to the metre. *Question 26*—SHORT LIST BOYS!—3, 6, 9, 12, 16, 29 or 36 sq ft in 1 sq yd ..."

The I.M.A. report on Interface I included a section on "Arithmetic in the Basic Training Workshop". I quote two passages from it to illustrate some of the skills needed but found lacking:

(ii) **"Addition and Subtraction**

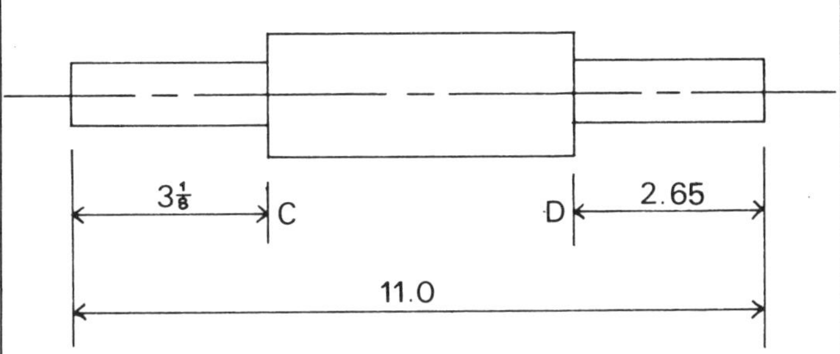

Fig. 1 All dimensions in inches. Tolerances: fractional $\pm\frac{1}{64}$; decimal ±0·005.

The most obvious example of the use of addition and subtraction is in dealing with dimensions. Fig 1 shows a simple stepped shaft with imperial dimensions. The shaft has fractional and decimal dimensions since this provides a convenient convention for specifying tolerances with differing degrees of "tightness". It may be necessary to find the distance from *C* to *D* during manufacture and this will involve a fraction into decimal conversion so that the appropriate addition/subtraction can be done."

(viii) **"Multiplication and Division**

Many simple work problems involve multiplication and division. For example, if a component is 50mm long including any cutting allowance, how much bar is required to make 500 components. Alternatively if material is to be removed from the bar shown in Fig 2 to reduce its diameter from 50mm to 46mm by how much should the cutting tool be advanced? This type of problem causes serious difficulty for some young people who cannot divide by 2. The trainees concerned are generally from the middle CSE range, and we only discover their problem when they start to produce faulty work. In fact a trainee could use any method for the solution of the problem and provided he produced an accurate job in the time allowed we would

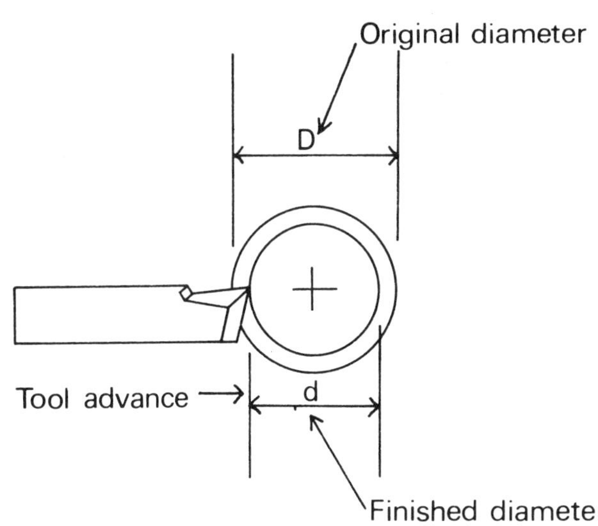

Figure 2

not question his arithmetic. The turning tool is mounted upon a tool post which is advanced by means of a calibrated handwheel. It would be possible to calibrate the handwheel so as to avoid the need to divide by 2, but this would cause other problems."

Interface II had a very different atmosphere from that of the first conference. It was clear that all parties had come to realise the complexity of the problem, and to appreciate the needs and difficulties of the "other side". Despite the greater harmony, on reflection I found the progress made disappointing. Mr R L Lindsay, the conference organiser, had made every effort to involve more industries, commerce, the nursing and medical professions, etc ... but with little success. There were virtually no school teachers there and the Examination Boards had declined to attend (Ref. 6). These were serious deficiencies for it was difficult to gauge the extent of the problems facing the employers, the syllabus constructors, and the classroom teachers. So what *did* happen?

A useful introduction was made by Mr J W G Boucher who spoke of the nature of Primary school mathematics and he made it clear that arithmetic is still a major priority of every teacher. Talk of the abandonment of learning tables is very misleading. There followed two talks, only marginally related to the main debate in my view, but nevertheless of interest. Firstly, Mr R A Parker, who was on a one year schoolmaster fellowship at Bath University, outlined the kinds of mathematical work done by "white collar workers" in a variety of British Engineering firms. His research was concentrated on employees with O Level in Mathematics, or equivalent, as a minimum. Mr Parker emphasised the growing importance of numerical methods throughout the industry, and also the elementary nature of much of the mathematics used, suggesting a wastage of talent or training. Secondly, Dr J A Anderson reported that so far (and his research is still at an early stage) he has found no startling significant differences between mathematics undergraduates at Nottingham University who had studied modern mathematics courses at school and those who had followed traditional courses.

A more relevant session was that introduced by Mr N G G Webb. He first reported on the uneven availability of computer facilities and then moved on to discuss the role of the pocket calculator. It soon became clear that many participants doubted that the pocket calculator would be common in most schools for a long time to come, despite some indications to the contrary. Again, many were sceptical of their value, anyway, as an aid to arithmetical competence, mental or practical. Indeed there was talk of calculators leading to atrophy of arithmetical skills, although this was disputed by some experienced teachers. What I had thought would be an important session petered out. "The lathe operator up to his elbows in oil can't turn to a pocket calculator every time he has to do a simple calculation." End of discussion. However, it would be quite wrong to dismiss so lightly the pocket calculator which is sure to have significance in both society generally and mathematics education.

The next session brought a lively report from Mrs R M Rees on research into the mathematical difficulties of students, being a useful continuation and extension of the work reported in *Mathematics in School* Vol. 3 No. 1 (Ref. 7).

The last main speaker, Dr A T Rogerson, did much to reassure those delegates who felt that SMP have a "take it or leave it" attitude. The 1974 conference was probably a factor which led to SMP's brochure on Manipulative Skills (Ref. 8) and Dr Rogerson confirmed that SMP would be publishing in 1976 "drill booklets" containing additional exercises encouraging basic arithmetical and algebraic work to supplement the 1—5 and A—H texts. It was also made clear that the current SMP 7—13 Project was taking very seriously the problem of the development of adequate arithmetical skills.

Each talk was followed by a lively discussion session and the conference was rounded off by two further discussion periods aimed at reaching decisions as to further action to be taken. The call for a "Bullock Report" equivalent on Numeracy was mooted. It was also suggested that the O Level Mathematics Examination should include a compulsory arithmetical section (or paper) with a high mark necessary on this part to qualify for an overall pass. In view of the recent grading changes this seems a non-starter but the alternative of a separate "arithmetic" grading, similar to the oral grading in Modern Languages, has some attraction.

The difficulty of legislating for CSE was noted but it was felt that a similar scheme could be encouraged where appropriate. (Some co-operation between schools designing CSE syllabuses and local industry is already taking place (Ref. 4) and may be the most fruitful approach). Despite valiant efforts by the chairmen of both final discussion sessions, little agreement seemed to be reached as to what we, the conference members, could do. We all recognized the seriousness of the problem but also saw that at least in the short term the answer must lie at the *local* level. And this is where the individual teacher must come in.

Action

Find out for yourself what kind of mathematics is really needed by your school-leavers. Contact the local Industrial Training Officers, and local firms and businesses. Consult your local Mathematics Adviser. Help to organize meetings of local mathematics teachers and employers. Look critically at your mathematics syllabuses and textbooks. Are you preparing your pupils adequately for future employment? Why do you teach the topics you *do* teach? Try to get hold of entry tests in mathematics used by local commerce and industry and show these to your colleagues *and* to the pupils and find out why the firms use them and what results they are getting.

Reaction

Well, are we seeking to put the clock back? We are today realizing that not all labelled "modern" must be the best and mathematics education is entering a period of consolidation rather than innovation. It was perhaps appropriate that the clock on the wall of the conference seminar room went resolutely backwards throughout the conference period! This led to some participants wondering how often the clock would be showing the correct time. For example at a correct time of

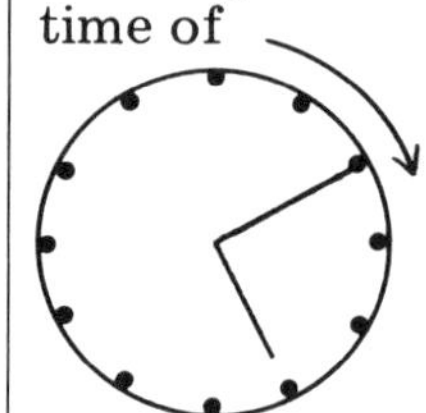

5.10

the clock showed

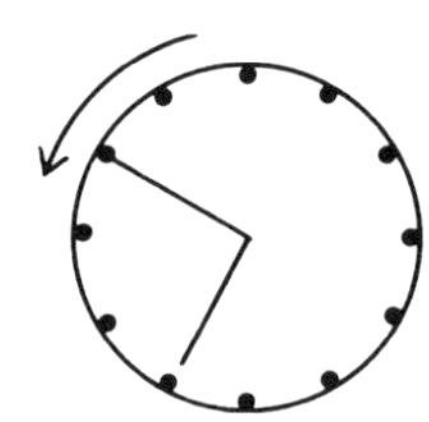

6.50

By a happy coincidence this problem can be solved equally well by arguments of symmetry or by more traditional methods! It would be cynical to say, as did one conference member, that the clock's behaviour reflected the "progress" we made—negative rather than positive—but it does remain true that little will come from our meeting unless individual schools take up the challenge. Those in industry are most anxious to help so why not meet them half way?

References

1. Mathematical needs of School leavers Entering Employment Symposium Proceedings *Series No. 6*. The Institute of Mathematics and its Applications, 1974.
2. Who needs arithmetic nowadays? A R Tammadge. *Mathematics in School*. Vol. 4 No. 1, January 1975.
3. Whatever happened to numeracy? R L Lindsay *Times Education Supplement: EXTRA*, p. 59. 4 October 1974.
4. *SKILL*, no. 16, Summer 1975. Published by the Engineering Industry Training Board, 140 Tottenham Court Road, London W1P 9LN.
5. *SKILL*, No. 15, New Year 1975.
6. The breakdown of membership of the conference was roughly as follows:

Teacher Trainers (Colleges and Universities) and Researchers	10
H.M.I.s and Local Advisers	5
F.E., Technical College and Armed Forces Lecturers	6
University and Polytechnic Lecturers	4
School teachers	2
SMP	1
Engineering Industry Representatives	5
British Petroleum	1

7. An Investigation of some common Mathematical Difficulties experienced by Students. Ruth Rees. *Mathematics In School*. Vol. 3 No. 1 January 1974.
8. *Manipulative Skills in School Mathematics*. SMP, 1974.

"IS IT AN 'ADD', MISS?"

Part I

by Margaret Brown and Dietmar Küchemann. Centre for Science Education, Chelsea College

The Concepts in Secondary Mathematics and Science (CSMS) project, financed by the SSRC, was set up at Chelsea College in October 1974, with the aim of investigating the levels of conceptual demand of various mathematical and scientific areas within the secondary curriculum. Having charted the parts of these subjects children find difficult, and tried to explain why, we hope eventually to be in a much better position to match topics and teaching approaches to the styles of thinking of individual children.

In the area of mathematics, we decided to start almost at the beginning and look at the understanding of the basic four operations on natural numbers among children mainly in the first year in secondary school, although looking briefly at children one year above and below this age.

Figure 1

1.
A bar of chocolate can be broken into 12 squares.
There are 3 squares in a row.
How do you work out how many rows there are?

12+3	3×4
12×3	3−12
6+6	12÷3
12−3	3÷12

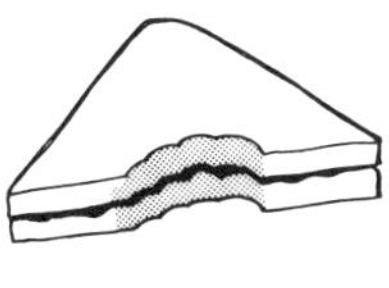

7.
A shop makes sandwiches.
You can choose from 3 sorts of bread and 6 sorts of filling.
How do you work out how many different sandwiches you could choose?

3×6	6−3
6+3	3−6
18÷3	6÷3
6×3	3+3

8.
A gardener has 391 daffodils.
These are to be planted in 23 flowerbeds.
Each flowerbed is to have the same number of daffodils.
How do you work out how many daffodils will be planted in each flowerbed?

391−23	23÷391
23−391	391×23
391+23	23+23
23×17	391÷23

We were interested not so much in computation as in whether children could see what operation should be applied in a given situation, and conversely whether given an arithmetic expression like 9×3 they could make up a "story" for which the expression was a model. (This second area is the subject of Part II of this article.)

We wrote a number of items like the three shown in Fig. 1, of which some involved small numbers (less than 10) and others larger ones (between 12 and 500).

We tried to vary the form of an operation involved; for instance the shop item above is more akin to the "Cartesian product" or "cross-product" form of multiplication (see *Computation and Structure 3, §4,* of the Nuffield Mathematics Project) than to the "repeated addition" form of multiplication which was embodied in the remaining two multiplication items in the test. Again the chocolate item above involves grouping (how many shares if they contain 3 each?) whereas the daffodil item is of the "sharing" type (how many each if they are shared between 23?)

We used these items to interview 35 children from all ability levels in the first and second year of the secondary school. The children were asked how they would tackle the problem, sometimes going as far as carrying out the steps if their strategy was not clear, and were only then asked to choose an expression from the sets shown on the right in Figure 1, which were on the reverse side of the problem cards.

Results of the Interviews

We classified the replies to the "problem" items into five different levels.

Stage 1: Intuitive. These children sometimes gave delightful idiosyncratic answers which showed they were playing the game on their own terms rather than on ours. Others, not knowing where to start, "guessed" wildly.

Daffodils

YG *You er . . . I know what to do but I can't say it. . . .*

MB *Yes, well you do it then. Can you do it?*

YG *Those are daffodils and these are flowerbeds, large you see . . . Oh! they're being planted in different flowerbeds, you'd have to put them in groups . . .*

MB *Yes, how many would you have in each group? What*

would you do with 23 and 391, if you had to find out?
YG *See if I had them. I'd count them up . . . say I had 20 of each . . . I'd put 20 in that one, 20 in that one . . .*
MB *Suppose you had some left over at the end when you've got to 23 flower beds?*
YG *I'd plant them in a pot(!!)*

Daffodils

DPi *About 200*
MB *About 200. How did you work it out?*
DPi *Just guessed.*
MB *Just guessed. O.K. which of those did you do?*
DPi *No.*
MB *No. O.K.*

Stage 2A: Early Concrete. Children were able to produce successful strategies for most, if not all, the items but these were always based on addition or subtraction.

Daffodils

MB *(Reads question.) What would you do with those two numbers to work it out?*
SB *23 there take away 23, 23, 23 . . .*
MB *Keep on taking away 23?*
SB *Yes.*
MB *Can you think of any of those that would be right? (Back)*
SB *That one.*
MB *23 take away 391. Any others?*
SB *That one.*
MB *391 take away 23.*

Stage 2A/B: Middle Concrete. Multiplication and division were sometimes recognised, but there were some inconsistencies. The sandwich item in particular was likely to cause trouble.

Chocolate

MBa *(Reads question to himself) 4?*
DK *Mmm . . . how did you work that out?*
MBa *3 times 4.*
DK *3 times 4 is what?*
MBa *12.*
DK *I see, how did you realise it was 4?*
MBa *Just worked it out in tables.*
DK *I see O.K. Which of these? (Back).*
MBa *3 4's.*
DK *Would any of the others do?*
MBa *Yes, 12 shared between 4 . . .*
DK *Why would that one work?*
MBa *Cos it's the same . . . that works out as 12 . . . so does that, so it's the same.*
DK *Well . . .*
MBa *4 shared between 12 works out 3 and so does that . . .*
DK *I see, what about this one though, 12 divided by 3?*
MBa *No . . . it only goes 3 times . . . that's 3, 6, 9, 12, that's right innit? No that's 4! Yeh that one works as well.*
DK *O.K. So we've got three versions here now, let's just sort of pin it down a bit more . . . if we look at what we started with; 12 squares altogether, 3 squares in a row, O.K. that's what you're told, which of those three would you think the better one?*
MBa *3 times 4.*

Stage 2B: Late Concrete. Fairly clear identification of multiplication and division items, although often the inverse was chosen, e.g. 3×4 for the chocolate item. No distinction between 391÷23 and 23÷391 was made.

Chocolate

MB *(Reads question).*
JN *4.*
MB *What did you do with the 12 and the 3.*
JN *Divided it.*
MB *(Back).*
JN *3 divided by 12 and 12 divided by 3.*
MB *Are they the same?*
JN *Yes.*

Figure 2: Question Facilities (Pupils aged 11-12)

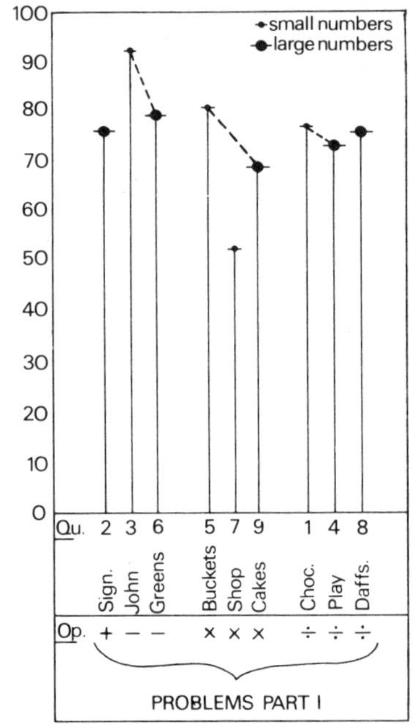

Stage 3A: Early Formal. Generally very quick answers to all questions, and a recognition of the non-commutativity of subtraction and division.

Daffodils

LD *How many 23's go into 391.*
BJ *Which one of those would . . .?*
LD *This one.*
BJ *That's right, now what does that one say?*
LD *391 divided by 23.*
BJ *Can we use that one? (23÷391).*
LD *No, because that can't go into that.*
BJ *That's right.*

One general point that we noticed throughout the interviewing was that almost all children could produce successful strategies for solving problems, even when they did not recognise the operations involved. Complementary addition (subtraction by "adding on" in the "shopkeeper's method") was successfully used mentally with large numbers by many children who could not recall the official subtraction algorithm. Also various, sometimes ingenious, forms of repeated addition and subtraction were employed by many who, although recognising that the problem required a multiplication or division operation, confessed that they could not remember the "rules".

There would therefore seem to be an argument for considering whether the traditional algorithms are the most useful ones and whether a method more closely related to children's "concrete" methods might not be easier to recall, even if slightly less efficient.

After interviewing we then tried out various versions of group tests which, although they clearly gave less information on any individual child than the interviews, did enable a whole class to be tested at once. Since the very straightforward items on addition and those on the "take away" form of subtraction were done correctly by almost all children, we omitted these and arrived at a final format which included, in Part 1, three problem items on harder forms of addition and subtraction, three problems on multiplication and three on division. (In Part 2 were the "story" items on subtraction, multiplication and division which will be discussed in the second part of this article.) In each part there was a trial item for discussion, which included revision of the operation symbols, and the teachers were also asked to take classes through the test together, reading all the items out twice, to avoid the difficulties with poor readers.

Results were gathered on 1,138 children in the first

year secondary (or equivalent), together with 147 in the previous year and 293 in the following year. The children came from different parts of England and Scotland, and from different types of schools, including primary, middle, secondary modern, grammar and comprehensive.

Results on the items

The item facilities, i.e. the percentages of first year secondary children who got each question correct, were as shown in Fig. 2. (The item was marked correct provided a child recognised a correct strategy, so that for example for the chocolate item the answers 12÷3, 3÷12 and 3×4 were all accepted.)

The items which are linked are those which were parallel in form except that one contained small numbers (indicated by small dots) and the other large. We judged that these differences, which are usually around 10%, were due in each case to those children whose grasp of the operation was rather tentative, so that large numbers pushed it beyond their threshold.

Comparison between the operations was difficult since some items represented very straightforward forms (e.g. buckets, cakes, daffodils) whereas all the others were chosen to represent more complex forms. Also whereas there was little difference between the two forms of division, with multiplication the "Cartesian product" item (shop) was significantly harder than the "repeated addition" form. In the story items which will form the subject of Part II the order was very clearly +, −, ÷, × in increasing order of difficulty; the results of Part I although more confused do not contradict this.

We tried to match the performance of children in the class test to the levels on the interviews which have been described above, and although the correspondence is rather tenuous we arrived at the table below. (The percentages are likely to be correct only to within about 4% for the 11-12 sample, within 7% for the 12-13 sample, and within 9% for the 10-11 sample.)

	Intuitive	Concrete Operations			Formal Operations
Stage	1	2A	2A/B	2B	3A (or above)
Age					
10-11	2% (7%)	23% (14%)	38% (36%)	31% (37%)	5% (6%)
11-12	2% (4%)	15% (8%)	40% (24%)	35% (56%)	7% (8%)
12-13	1% (3%)	12% (6%)	30% (21%)	47% (56%)	11% (14%)

The percentage of children recognising the difference between 84÷28 and 28÷84 was rather more than those in the 3A category; the total at 10-11 for instance was around 18% while that at 11-12 was 20%, and at 12-13, 25%. However, not all these children had a firm grasp of multiplication.

The figures in brackets are the percentages estimated to be in the corresponding Piagetian stage by the science team of CSMS, using a class form of an original Piagetian task concerning volume and density, with 5,000 children. In the assessments of both our test and the science tasks it is particularly difficult to distinguish where the middle sub-stage of concrete operations begins and ends, and the two sets of figures more nearly correspond if the three substages of concrete operations are taken together.

In Part II of this article we will discuss the results in the "story items" together with some of the apparent reasons for the difficulty of the concept of multiplication.

Editorial

Volume 5 No 2 March 1976

The need for a numeracy report

Without doubt, the Bullock Report on literacy has made a big impact in schools. It has highlighted a very serious situation which is now being tackled by courses, conferences and other appropriate action. Do we need a similar report on numeracy? The evidence would seem to indicate that we do.

One of the major factors in the public concern about numeracy is that there are more jobs which require mathematics. Thus, engineering employers are taking apprentices much further down the ability scale, and their lack of numerical knowledge is showing more clearly. When, in 1963, the Newsom committee reported on *Half Our Future,* it was stated that numeracy was a problem without much hope of a satisfactory resolution. To quote from paragraph 335: "For the least able it will be sufficient if they can apply the arithmetical rules which average children have mastered by the end of the junior school, with confidence and understanding . . .".

The same less able boys and girls are now joining firms on apprenticeships, and some of their employers' comments are withering. To quote again, this time from the Engineering Industry Training Board's *Blueprint:* "Industry cannot afford to spend time educating trainees in the basic number work and schools must be forced to re-look at their objectives and educate to needs of industry which surely cannot be too removed from those of society as a whole".

To be fair, other articles in the same issue refer to the serious shortage of qualified teachers, and to the fact that the tests used by firms are often kept secret from teachers (sometimes commercial test suppliers enforce secrecy on users) which, of course, does not help teachers to know what industry wants. There's also a sting: ". . . the 'remedial maths' provided by firms and groups and colleges appears to be highly effective — and always 'trad', not 'modern' maths".

There are other important factors of which teachers need to be made aware. It's very easy to criticise the engineering industry as a whole, and it is true that the big companies could, sometimes, be more enlightened in their approach. But, can we reasonably expect the 13,000 small firms which employ less than 25 people to have the capacity to educate themselves? Do they have a right to expect that schools will provide readymade employees? Should the education system attempt to speak to them? Do teachers realise that it's not always possible to use pocket calculators in dirty wet conditions, and that often mental calculations are much quicker than cleaning oily hands?

A lot of people are blaming each other in what to many must appear to be an unseemly fashion. It's not easy to be constructive but we hope that the Association will play its part in reaching a satisfactory outcome. If a Bullock type report is proved necessary, it will be up to all of us to honestly expose our prejudices and weaknesses for the total good of the community.

Fifteen Year

Mathematics in School 1960-75

In the late fifties and early sixties, life was comparatively simple for the school mathematics teacher. Most teachers followed a well-established, fairly rigid syllabus, probably the same one they had themselves experienced as pupils. Few had heard of Piaget, "modern" mathematics, mixed ability groups or "discovery" learning. In most schools mathematics lessons were periods of instruction rather than education. The teacher could be lazy; external examinations followed a set pattern and it was possible to drill pupils by doing a large number of examples from past papers. The mathematics teacher was able to complete his "Times" crossword during lessons each morning and yet still achieve excellent examination results.

Murmurs that strange things were happening to mathematics in America reached the teacher from time to time. The conference at Southampton in 1961 attracted considerable publicity. In 1962 the Mathematical Association produced a booklet, "Modernizing School Mathematics" and slowly the teacher in this country was finding it hard not to be aware at least of the fact that change was in the air. There was a conference arranged by Shell in London in 1963 and memories are of free drink, a bar that never closed, good food and an energetic Bryan Thwaites rushing importantly around, but above all of a feeling of tremendous enthusiasm among teachers for change in school mathematics. Before all this there had been Geoffrey Matthews addressing the 1962 AGM on "Matrices for the Million" and suggesting the revolutionary idea that we might teach matrices in junior secondary school classes. An issue of the *Gazette* in February 1963 suggested deliciously new, attractive and interesting material for both teacher and pupil: the novelty would take some time to wear thin. The "spotty" St. Dunstan's booklets began to appear, they were new, fairly cheap and easy to use, above all they were different from what was already available. The ATM's "Some lessons in Mathematics" arrived in 1964, a remarkable and influential book with a wealth of ideas.

Resistance to change was strong. In one school the head of mathematics had some SMP books which were kept securely locked in a cupboard; it was hard for the young teacher to get near them, impossible for the pupil. But many teachers were eager to use new material and there are numerous mischievous stories of senior mathematics teachers unable to help classes with work on "Sets" or "Binary numbers" when they deputized for young absent members of staff.

The name "Nuffield" was heard from junior schools, draft guides were produced and it was not to be long before secondary schools would have to take notice. Many other projects began to appear, new textbooks arrived fast, it was impossible to be unaware of change, hard to resist pressures to adopt new syllabuses. Strongly influenced by the Nuffield attack, there now began a move from formal class teaching towards pupil *activities* and phrases like "discovery learning" and "open-ended question" were rapidly becoming part of the teacher's vocabulary. For the first time for many years it seemed, teachers were forced to ask themselves the question, "Why teach mathematics?".

There were strong moves to place much more emphasis on *applications* of mathematics. Calculating machines, once a rare and odd curiosity, became familiar sights in many classrooms. The "Computer" was now powerful and important enough to merit discussion in school although there must have been many fruitless visits to local computer installations where groups of pupils would stand and gaze in excited anticipation and awe at the strange monster while the "expert" breathlessly reeled off his long, incomprehensible description. The phrase "in-service training" appeared frequently, there were many courses for teachers and, thanks largely to Nuffield in some areas, teachers' centres were created with amazing speed. CSE arrived with freedom for the teacher (if he wanted it) and still more books, apparatus, projects, workcards in ever growing numbers.

By the second half of the sixties many teachers felt that change in school mathematics had become like a driverless steamroller that many could cling to but no-one could stop. There were powerful critics of the "new" mathematics, notably Hammersley and Lyness while other countries were not without their problems:

*"A 35 year old French mathematics teacher last year committed suicide because, his family allege, he felt that neither he nor his pupils could cope with modern mathematics." (*Times Educational Supplement, *February 1972).*

"According to Dr. James Shackelford, an environmental chemist with the US government, New Maths makes cruel and unreasonable demands on kids by stressing abstruse mathematics

s of Change

by M. L. Cornelius, Department of Education, University of Durham

theory very few can understand at the expense of the ability to perform. Children learn at a pretty well fixed rate, and there's almost nothing of which children this age (under 11) can understand the theory. You've got to say, This is the way it is kid, and you'd better learn it—by doing, the way you learn to tie your laces or drive a car." (Times Educational Supplement, *February 1973).*

But now the sweeping changes had gone too far and gathered too much momentum to be stopped.

Changes in schools, the curriculum and mathematics all made contributions towards the creation of new areas for discussion like team teaching, mixed ability groups, workcards, mathematics for the slow learner, ROSLA, Mode 3 syllabuses, computer terminals . . . In many situations something like mixed ability teaching in mathematics has probably arrived as the direct result of the adoption of new courses but the teacher has often been faced with tremendous change simultaneously in both subject content and method of teaching.

In 1960 we were desperately short of mathematics teachers but life was fairly simple and pupils were made to work hard. Now, fifteen years later, we are well on the way to having teachers who think more about their teaching, attend ccurses and teachers' centres, work much harder and are generally far better informed than their predecessors. But while the teachers work harder the pupils do less. Too easy now for the lazy child to prolong a simple project or workcard or to leave most of the work to someone else in the group. Too easy for the idle pupil to drift along in the "interesting" lessons without bothering about tables, manipulation or technique. Too easy to avoid the hard work when it comes along. Have we succeeded in producing a thinking, well-educated teacher only to find we now have pupils who lack persistence, technique and much important mathematical knowledge? Our pupils can now happily "play" with a topic for hours but do they know and enjoy mathematics any more than our pupils of fifteen years ago?

The prospect for the second half of the seventies is gloomy. There are many areas where mathematics in school is facing difficult problems:

(i) **The Computer** has become a threat to mathematics teaching. Teachers are expected to spend more and more time on "computing" frequently at the expense of mathematics. Fine for the enthusiast but there must be many mathematicians who are not interested in teaching about the computer. Unless "computing" becomes a subject in its own right, mathematics teachers will find increasing demands on their time in the future.

(ii) **"Mixed Ability" teaching** can have many different meanings but the idea is causing much damage to mathematics teaching. Perhaps mixed ability groups work in junior forms, they certainly will not work for ever. Where should we stop? First form, third form, . . . sixth form? There are doubtless many successful enthusiasts but pressures are building up for the more reluctant teacher:

"It occurs to me that in the near future, several teachers—whether they choose to or not—may be forced to undergo a . . . transition (to mixed ability teaching)." (Mathematics in School, *July 1974).*

So much for the freedom of the teacher.

(iii) **Multiple Choice Questions** are a great convenience for the examiner but threaten to trivialize examinations in mathematics. Is mathematics all about problem solving or is it picking answers with a pin and rarely having to write more than one line of explanation?

(iv) **Books, Apparatus** and other new materials reach the confused teacher in ever-increasing numbers. Are there too many new books? Too many projects? Too many new gimmicks in school mathematics?

(v) **Calculating Machines** are now easily available and becoming cheaper. Some teachers use the existence of these machines as an excuse for not teaching basic arithmetic. The day when every pupil has his own calculator is still a long way off, indeed one hopes it will never arrive! Perhaps machines do not make mistakes but it is easy to press wrong keys, feeding in incorrect information.

(vi) **Other subjects** like Geography, Economics, Social Sciences . . . are making increasing use of mathematics. It is good to see such uses being made but again the pressures on the mathematics teacher will increase.

There are many problems ahead for school mathematics, most can be overcome. It would be comforting to know that in 1980 pupils will work as hard as their predecessors in 1960. ■

SMP a disaster?

by D. M. Burley, Dept. of Applied Mathematics and Computing Science, The University, Sheffield

The development of SMP courses in schools over the past fifteen years and its apparent popularity with teachers has led to a crisis situation in mathematics, particularly in the applications of mathematics. Already warning signals have been fired that this view is not out of place in the thinking of some mathematicians, see the *IMA Bulletin*, August 1973 and the editorial comment in *Mathematical Gazette*, June 1975. The present article takes a slightly different line from the editorial and concentrates more on the service aspect of the subject and one specific scheme of study.

There seem to be three main groups of children in schools (i) those gifted at mathematics (ii) those who are unlikely to do any more mathematics after 16 (iii) those who wish to use their mathematics in physics, chemistry, etc. The first group is a small one about which there is little problem since they will survive most schemes of work and most teaching methods. The second group consists of 75 per cent or more of the school population. The main concern with this group is that they can perform efficiently the usual arithmetical operations and have simple concepts like area, volume, capacity well established. Hopefully most of these ideas are known by the time a child is 11 years old so that the mathematics taught to this group in the secondary school is less crucial to future studies in other scientific areas. It must, of course, be correct mathematics and it is a necessary part of the general education of a child. The distinctive method of thinking in mathematics should be of considerable value in later life and an SMP syllabus seems satisfactory from this viewpoint.

My major concern here is about group (iii) which is taken from the top 20 per cent of the ability range of a school population. From this group will emerge the country's engineers, metallurgists, technologists, chemists, physicists, economists and biologists. These are the people who will be the prime producers of the wealth of the country and it is just this group that is badly served by the SMP books.

All the damage is done in the early secondary years and therefore it is the "O" level syllabus about which I take issue. I have little complaint about the content of the A-level syllabus which seems to be perfectly satisfactory. My only criticism of it is a general one of the whole scheme that there is insufficient practice given through simple examples. My major concern is therefore the 11-16 age group. The foundation on which all the science and technology based subjects stand is a good grounding in physics and chemistry. As a child progresses from the purely observational stage in the sciences to a quantification of these observations, it is mathematical skill and awareness that provides the link between the two. The process involves the production of a mathematical model to give a clearer understanding of the physical situation. This approach has been a standard technique in physics since Newton's time. It has developed through chemistry and the technologies and, comparatively recently, into economics and biology. It is clear that the system cannot create gifted mathematicians but it can and should teach the technical skill and understanding to the capable group of people under consideration to enable them to cope with these simple mathematical models. Usually, the group's prime interest is not mathematics for its own sake but in what mathematics can do for them in their own subject. Mathematicians should not, as I think SMP has done by deliberate intent or, more likely, by oversight, dismiss the service aspect of their subject and they should not regard mathematics in an egocentric way as a subject only for the elite. It is here that I think that SMP has done a great disservice to the community.

Expanding my thesis in a little more detail, it is the arithmetic and algebraic skill which is not encouraged by SMP. A secondary school teacher can reasonably complain that arithmetic skill should be there by the age of eleven. With this I would heartily agree. I am, however, pessimistic enough to believe that in many cases the necessary skill is not there and that primary school so called "modern mathematics" has a lot to answer for. I have little experience with the primary area but it appears to me that "modern mathematics" in primary schools is an easy option siezed upon by many teachers who have a sketchy knowledge of mathematics and no confidence in where their mathematics is leading them. At this level, as at all other levels in mathematics, there is no substitute for good, solid, hard work with lots of practice at, now unfashionable, pages of sums. I have digressed from my main theme however, which concerns work in 11-16 age group. Having accepted that arithmetic may not be as well established as it should be, the SMP syllabus does not encourage and continually revise arithmetic skills. It is the duty of every teacher to assess these needs and to supply remedial material to supplement the texts.

Just as important as arithmetic is some mastery of elementary algebraic operations. It is here that a young person first meets problems in quantification of physical ideas and their translation into algebraic entities. I quote examples from physics since this is the area I know best and where the problems are most acute. Take for instance an experiment that illustrates Ohm's law. Through a given resistance a known current is passed and the voltage drop is measured. By a succession of such experiments it is soon discovered that

$$V = IR$$

to a good approximation. Having established this fact, the next step is to use the relationship in different contexts to predict the behaviour of a system. It is here that the SMP trained student suffers since he finds the manipulation of the Ohm's law formula impossibly difficult. Instead of an application of "cross multiplying", a technique of universal use, the student will go through several laborious steps to achieve for instance

$$I = V/R.$$

Even the simple process of cancelling out the Rs in $(IR)/R$ is not an automatic reaction. Thus has started the first nail in the coffin of physics as a viable subject.

The next experiment might be to measure the coefficient of expansion of a bar and an attempt to obtain

$$l = l_0(1+\alpha T).$$

Here the major problem is to identify α from the other data. The first difficulty with this relation is the "dreaded brackets"; fortunately the expression is not compounded with the even more "dreaded minus minus". A SMP trained person would struggle through four or five tortuous operations to obtain

$$\alpha = \frac{1}{l_0 T}(l-l_0),$$

and, I suspect, would never actually get there since he would have to voluntarily insert a bracket. The problem is the simple manipulation of an algebraic expression, which, by the third form in a secondary school, should be second nature to the group of children under consideration. However, a second nail has been embedded in the coffin and the lid is slowly closing.

A final example that finishes off any hope of redemption is the well known lens formula

$$\frac{1}{f} = \frac{1}{u}+\frac{1}{v}.$$

In this expression the SMP trained person will tell you quite happily that division and addition do not commute but is likely to proceed as follows:

given $u = 25$, $v = 15$

$$\frac{1}{f} = \frac{1}{25}+\frac{1}{15} = \frac{2}{40} = \frac{1}{20}$$

$$\Rightarrow f = 20.$$

(*Note:* this is the exact calculation of one of the better students quoted to me by a physics teacher.)

Having failed to cope with any ease with these three examples, and I'm sure the reader can supply another dozen or so similar cases, who can blame a child for giving the subject up as a bad job? Who wants to continue a course of study which is a continual battle against insurmountable odds? I'm sure that I can hear the comment that physics and chemistry teachers have always complained that their pupils are never taught the right sort of mathematics at the right time. This I accept but with the sort of algebraic training given in SMP the complaint must be an order of magnitude stronger.

The next question to be asked is what can a teacher do to remedy the situation? Here, I think, it rests fairly and squarely on the shoulders of the head of the mathematics department of a school. He is given and paid for the responsibility of thinking about the syllabuses taught in his school. He must therefore insist that his staff do not slavishly follow every word of the book and never deviate from the gospel according to SMP. To the group that I have been considering, that is the capable sets in the top 20 per cent in academic ability, additional material should be supplied that corrects the obvious weakness of SMP on algebraic manipulation. This material should include enough work on cancellation, cross multiplying, brackets, minus-minus, rearrangement of equations to make the processes more or less second nature. There appears to be a marked reluctance to set examples on what used to be called "mechanical" mathematics but for me it is exactly what is needed, even if it involves setting pages of "sums". I think it quite possible to dress up such pages to make them acceptable to the modern student or perhaps more significantly to the educational psychologist. What many mathematicians seem to forget is that mathematics is largely about thinking hard and carefully about a class of problems, developing general algorithms for solving such problems so that the process becomes a purely mechanical one. This relieves the thinking process for the next class of problems.

The reader may be excused if he has read into the previous paragraphs a criticism of all that is in SMP. Quite the reverse, the work that is done on matrices, vectors and statistics is, in my opinion, admirable and several other topics are covered perfectly adequately. I am, however, less happy about the rather obsessive use of set theory and set theory notation and I am fully in agreement with the comments of Hammersley (1968) and the *Mathematical Gazette* editorial (1975). It is not necessary to get involved in playing with the notation, which is all that set theory involves at this stage, and the formal manipulation of sets is best left until a student reaches University level when the subject becomes useful and relevant. At an informal level set theory is used continually and in schools an informal approach is perfectly satisfactory.

There have been revisions of the SMP books and additional cards are available but these go only a small way to remedy the situation. Many schools will be unable to afford the latest issues and will have to be content with their original versions until they actually fall to pieces. Consequently my comments appear to be applicable for many years to come.

Finally I would like to summarize this article. The country depends on high level science and technology for its existence; it cannot afford the luxury of the majority of the population studying the culture of this and other countries to the exclusion of the scientific disciplines. One of the major planks of high technology is elementary physics, which in turn depends on competent mathematical manipulation of simple expressions. SMP fails to give this mathematical training at an early level and students are rapidly turned away from physics and driven to study subjects less productive to the economy. It is the duty of every mathematics teacher to reconsider the service element of his teaching and to remedy the SMP disaster! ■

References

IMA Bulletin, 9, 1973, 238-252.
Math. Gazette, 59, 1975, 61-64.
J. M. Hammersley, *IMA Bulletin*, 4, 1968, 66-85.

SMP a disaster diverted?

by John Hersee

In an article in the last issue of *Mathematics in School* "SMP—a disaster?", Dr. Burley of Sheffield University expressed his concern for school mathematics courses and drew our attention to some very definite aims for mathematics teaching. It is good to see such interest, and I share Dr. Burley's concern for our future scientists, engineers and technologists. I am concerned, too, for the other two groups of pupils that he mentions —both the mathematically able and the majority who may make little use of their mathematics after they leave school. But I don't think that these three sets are non-overlapping, nor precisely identifiable in the secondary school, and, leaving aside the depressing overtones of the word, I am not sure that the most able *will* "survive" a bad course and poor teaching.

It would be interesting to discuss with Dr. Burley the various points he raises, but I doubt whether much would be achieved by a line-by-line reply to his article. Sufficient to say that the SMP O-level *syllabus*[1] lists a lot of arithmetic of the kind to be found in any traditional O-level syllabus, although logarithms are not included and algebra of the traditional kind is much less prominent. It is also worth mentioning that the SMP has issued a booklet on manipulative skills[2], setting out its views on what are reasonable and appropriate targets at various ages. While there may be few obvious sets of arithmetical examples in the SMP text-books (there are more in the "lettered" books than in the "numbered" ones), many of the topics in the course encourage and require arithmetical proficiency. Work on functions, matrix manipulations, probability, statistical calculations, for example, all involve considerable use of the basic skills.

Dr. Burley's remarks on algebra are more worrying. "Cross-multiplying" is a term which most teachers of mathematics avoid, along with other inefficient and ineffective rote-learned rules of thumb, and I wonder how many steps Dr. Burley requires to obtain $\alpha = \frac{1}{l_0 T}(l - l_0)$ from $l = l_0(1 + \alpha T)$; as for $\frac{1}{v} + \frac{1}{u} \Rightarrow \frac{1}{f}$, that must be one of the most quoted horror stories in physics since time immemorial!

We must realize that many factors interact in school mathematics—the syllabus—the text-book—the teacher —the class—besides external examinations and pupils' possible careers. I agree with Dr. Burley that it is the responsibility of the Head of Department in a school, and of every teacher of mathematics, to assess the mathematical needs of his pupils and see that these needs are met. Sufficient practice in routine calculations and manipulation must be provided, and, if the pupils need it, the text-book's treatment of a topic, such as algebra, must be supplemented.

But the assessing of pupils' needs in mathematics was a major element in the creation of the SMP—and of other projects. It was felt that the courses in use in schools were not suited to the needs of all pupils. One aim was to balance the emphasis on manipulative technique with an *understanding* of the underlying mathematics. I would agree that it is necessary for most pupils to be able to "perform efficiently the usual arithmetical operations, and have simple concepts like area, volume, capacity well established"—*necessary*, but not *sufficient*. I hope we can agree that the apparent dichotomy between understanding and manipulative facility is a false one; neither can be successfully mastered without the other.

There were other influences for change in the content of school syllabuses. School mathematics had become increasingly out of touch both with university mathematics and the mathematics which was (and is!) being applied and used in an ever-widening range of subjects and situations; the mathematics shaping our society. Dr. Burley's concern for the "users" of mathematics re-emphasises this aspect. And if we are concerned, for the moment, with those "taken from the top 20% of the ability range" who become engineers, scientists, and so on, among them will be the future leaders in these fields; those who in new work must draw on their mathematical background. They will need a sound mathematical background, chosen both for its present relevance and for its potential in developing a good attitute to new applications; in short for encouraging *adaptability* in a changing world: If the mathematics of school had remained unchanged we would today indeed be heading towards disaster, in a situation where the separation between school mathematics and the future mathematics and the future mathematical needs of our society would be ever increasing. Not all that is new is necessarily good, and a balance must be found between new and old; a balance which the teacher must strike.

We can now see that curriculum reform and development in mathematics is not brought about by a simple, if radical, change. It is a continuing process which must react, both to the degree of success of earlier changes and to the changes in society and in the world of mathematics in its widest sense. Many people can contribute to this continuing process through their interest in school mathematics. Those, like Dr. Burley, in universities, polytechnics, colleges, and those who employ school-leavers, have much to contribute from their respective backgrounds, and their comments and critisms are the more valuable when based on knowledge both of their own field and of the school courses

continued opposite

of today. A study of text-books and syllabuses makes a good starting point for the latter, while the Mathematical Association journals[3] and Reports[4] can help to complete the picture. For teachers, who see most easily the situation in their school, information and comment from those outside indicate matters to be considered and lines for future development. The debate continues!

References.

1. The School Mathematics Project. Syllabuses for the SMP GCE Examinations.
2. The School Mathematics Project. Manipulative Skills in School Mathematics.

   Copies of both of the above can be obtained from: The SMP Office, Westfield College, Hampstead, London, NW3 7ST.
3. *Mathematics in School* and *The Mathematical Gazette.*
4. See especially, for example *Mathematics 11-16* and *Primary Mathematics—a further Report,* both obtainable from The Mathematical Association, 259 London Road, Leicester, LE2 3HE.

John Hersee becomes Executive Director of the SMP this month.

BOOK REVIEWS

Mathematics for Schools — Level II, Books 1-6

Senior author: Harold Fletcher, with Arnold Howell for Books 5 and 6. Assistant author: Ruth Walker; a team of English and Canadian authors contributed. Addison-Wesley.

Harold Fletcher died before this project, for ten children's books covering the ages 7-13, together with Teachers' Resource Books, could be completed, but the remaining authors have drawn largely on his inspiration and planning of the series, which is an enduring memory to his work and influence on mathematics in Primary schools. Books 1-6 have now appeared, and take the work up to about the ten-year-old stage for many children.

The spirit of the series is modern, and is committed to discovery and understanding before technique, although consolidation of technique is not forgotten, for, as the authors say: "It does not follow that children always remember what they discover."

The main topics covered are number, measurement, space and shape, statistics and probability, and algebraic relations. The chief emphasis is on number and its properties, following the definition in the introduction to each Resource Book, "Mathematics is concerned with structural relations between concepts of number and the application of those concepts to environmental situations." One of the strongest features of the series is the extremely careful building up of children's understanding of concepts involving whole numbers and fractions, including decimal fractions. The pace is unhurried, the numbers used are always within the children's ability to illustrate with apparatus, and children are expected to develop their own methods of recording and only gradually to adapt these to conventional methods. Many teachers are looking for the support of a series of books such as this, to help them to structure the children's experience in such a way as to ensure real progression in understanding.

Much of the success that an individual teacher will have with the series will depend on her skill in ensuring that children have an adequate variety of preliminary and follow-up activities, in addition to working through the pages of the books. In fact, the suggestions for additional activities in the Teacher's Resource books form an essential part of the work, and themselves need supplementing. The weakness of any series of books is that it cannot make adequate use of the environment of a particular school or of the individual interests of children and teachers. This series is no exception, but certainly if it is used as the basis for the structural part of the work, and is supplemented by a good deal of environmental work and support from other parts of the curriculum such as science, it can form the basis for a very strong course.

There are several interesting features, which result from the authors' determination to use a "modern" mathematical approach, so that when discussing mathematical ideas with the teacher, the authors "have attempted to make our discussions informal in nature (with illustrations to clarify the context), but specific and accurate in content and definition." In this area of their work, the authors have been rather less successful than in the children's work. For example, "mass" has been correctly used throughout, and wording such as that given below elegantly circumvents, in the children's books, most of the difficulties of the traditional wording about weighing:

"Tim made these masses from 10 gramme masses of sand.

Which of the four masses will balance the parcels?"

The explanation for the teacher is less happy:

> " 'Mass' is the quantity or ***bulk*** of matter. On earth, 3kg of sugar have three times the bulk or mass of 1kg of sugar. On the other hand, whereas 1kg of sugar has the same *mass* on the earth as on the moon, its ***weight*** is not the same."

Is confusion between mass and weight to be replaced by a new confusion between mass and volume, for is this not what "bulk" is?

The series is fully, and on the whole successfully metricated, although the odometer reading of the car with registration number UMT 985K is given as 7892 km, which stretches present-day realism slightly. The authors have not, however, done the rethinking of work on fractions and decimals which the introduction of decimal notation at an early stage, as in £1.13 and 1.26m, makes possible. Indeed, the children multiply fractions in Book 6 before they are introduced to hundredths in any general context. The reviewer hoped that one of the consequences of decimalisation and metrication would be the postponement of abstract work with fractions until a later stage, and that decimal fractions would be used to a greater extent in children's practical work, as is appropriate at this stage. In fact, the major example of a difficult abstract argument in the series so far is that on equivalent fractions in Book 5. The reasoning (page 22)

I know $\frac{2}{2}$ is a member of the family of 1,
so $\frac{1}{2} \times \frac{2}{2} = \frac{1}{2}$
I also know $\frac{2}{4} = \frac{1}{2}$, so $\frac{1}{2} \times \frac{2}{2} = \frac{2}{4}$
$\frac{1}{2} \times \frac{2}{2} = \frac{1 \times 2}{2 \times 2} = \frac{2}{4}$

does not seem very likely to convince many ten-year-olds.

One of the features of the Teacher's Resource Books is the Glossary of mathematical terms. Here, insufficient thought has been given to problems of circularity and of helpfulness to those teachers whose knowledge of modern mathematical terminology is insufficient to enable them to read the text easily. For example, to clarify the problem of equivalent fractions, we consult the Resource Book for Book 5 (page 28) and read:

> "We can partition the set using the relation 'is a way of writing' . . . The relation 'is a way of writing' is an example of a special kind of relation known as an ***equivalence relation*** because it satisfies the reflexive, symmetric and transitive tests".

Nowhere in the glossary or the text can I find a hint of why a relation can sometimes be used to partition a set, and neither "reflexive" nor "symmetric" appears in the glossary, although reflection and symmetry do appear in their geometric uses, which may make the quoted sentences seem particularly nonsensical. This is a great pity, as the idea of a relation is one of the dominating themes of the work in the children's books, and the work on algebraic relations, in particular, is most skilfully developed, and forms an outstanding introduction to algebraic ideas at exactly the right level for the children.

The reviewer has a number of other doubts and queries about the mathematical exposition in the Resource Books. The teacher should not, however, be put off by these difficulties from making use of a series which has a number of excellent features, particularly its structured development of number and the excellent introductions to the new topics of probability, statistics and algebraic relations.

H. B. SHUARD

M=C+P: SOME THOUGHTS ON ATTAINMENT IN MATHEMATICS

by F. R. Watson, Institute of Education, University of Keele

Mathematical attainment is constantly in the news. Worries about numeracy, APU, proposals for tests of basic mathematical skills, DES surveys, "moans" from industry, even personal intervention at Prime Ministerial level, testify to the widespread interest and concern. After a "Minister for Drought", maybe a "Secretary of State for Numeracy"? A little cynicism may be forgiven, perhaps; all too often one sees complex and long-term problems elevated briefly to current issues of policy only to fade into obscurity when "instant" solutions make no impact on the situation. The numeracy issue is not new, as T. J. Fletcher[1] has pointed out. There is a danger that an ill-considered response to current pressures for an improved level of mathematical attainment may do more harm than good. For we have to examine the question, "What is mathematics in the curriculum *for*?" Why teach it at all? What contribution does it make to the education of the individual? And what do the "customers" — parents, employers, further and higher education, pupils (?!) really want from us?

So we return to the assertion in the title: "Mathematics equals conjecture plus proof". I want to argue that there is more, much more, to mathematics than numerical calculation. If we need to compute $12.75^2 + 5.92^2$ or $2\,468 \div 137$ there are powerful tools at hand to help us. Granted that we are not just concerned with button-pushing — pupils need to know whether to divide or to multiply, what is an appropriate degree of accuracy, a reasonable order of magnitude for an answer. Granted, too, that elementary mathematics is used in industry, in engineering, in science, geography and many other school "subjects", many of which demand (ideally) particular facilities and skills, sometimes geometrical, though largely numerical. Yet there is another aspect which may be in danger of being lost in the welter of they-need-to-know-how-to-do-this-calculation-so-never-mind-why-it-works-just-do-it. Mathematics is, or ought to be, the least arbitrary of subjects. $(a+1)^2 = a^2+2a+1$ *not* because the teacher says so, but because it can be demonstrated to be so. Pupils ought not to be asked to accept mathematical statements on the teacher's authority when they are capable of deciding for themselves whether they are true or false. In this example one could "try some numbers" — (which would dispose of $(a+1)^2 = a^2+1$); draw a picture of a square with a border; use the "rules" of multiplying-out brackets (again reasonably derived, not vouch-safed by some Moses-teacher as "this is what you do").

The usefulness of mathematics at higher levels is its ability to systematise knowledge and generate predictions. The *mathematics* of long division is in demonstrating *why the algorithm works* (which we, probably wisely, very seldom do!), not in performing it more slowly, less accurately and more tediously than machines can.

As I was writing this, a copy of *Mathematics Teaching*, No. 82 arrived. An article by Bob Jeffrey ("Making and Testing Conjectures", p. 15)[2] is a very apt illustration of what I have in mind. He describes some work with children using Colour factor rods to make single colour trains. They were asked to find the shortest trains of equal length for a given pair of colours. For example, with scarlet (4 cm) and violet (6 cm) the shortest equal trains have length 12 cm.

Children offered rules like "you just times the lengths", and later, when it becomes clear this doesn't work, "it doesn't work when the lengths are even", and "it doesn't work when you find a number that goes into both lengths".

Jeffrey comments, "Viewed from the content standpoint it could be argued that not much was achieved in these two sessions — perhaps a better feeling for factors and multiples is all that could be claimed. However, if we consider the activity as an experience of the process which leads to the establishment of mathematical results, we find many of the strategies which are typical of this process — generating and tabulating results, constructing classes, making conjectures, identifying new questions, testing conjectures, engaging in discussion and arguments about individual conjectures, and establishing agreed results."[2]

Of course, it would have been simpler to tell the children "its the LCM" and give a rule for working it out. But that's not the point. It is "giving rules" which leads to the sort of half-understanding where children muddle up area and perimeter — not, I suspect, *solely* because of a misunderstanding of words, but because "you have to find these two things and for one of them you add the numbers and double it, and for the other you times the numbers".

Wheeler asks, "Why do the majority of mathematics teachers not encourage their students to function as mathematicians, rather than try to lead them through a relatively restricted number of skills practised mainly in a routine fashion and which bear little obvious relation to mathematics as it exists beyond school? . . . The shift in emphasis wouldn't mean that students would not learn a good deal of the mathematics that is conventionally expected they will learn at school, but that this would be a by-product of the exercise of their own mathematical powers."[2]

The IMA test "revelations"[3] may increase the pressure on teachers — to concentrate on trying to ram home some corpus of basic knowledge which "every child should know", whether they understand it or not; to tell children, "Don't think, just *do* it", when we ought to tell them "Don't just do it, *think*!" Many of the people who are expressing concern about current problems would genuinely deny that they wish children to be *restricted* to a basic core — their concern is only that mastery of this "core" ought to be part of the teacher's goals. Yet this is a question of *priorities*, and if the transmission of basic skills becomes an over-riding objective, much harm may result.

Of course, there are mathematical skills which desirably children should have, but to make them the focus of our teaching, rather than the by-product Wheeler refers to, would be a tragic mistake, *because it would imply neglect of our prime aim* — the development of children's mathematical powers. What we *mean* by mathematical powers depends on what we understand by (doing) mathematics — hence my title, M = C + P. (Admittedly I am here defending a standpoint, and perhaps even overstating a case! Others will argue that mathematics = modelling (applying mathematics to real world situations) or M = abstraction and generalisation. I wouldn't want to deny these aspects, but *do* want to say that mathematics is more than the "arithmetic and equation-bashing" into which some teachers seem to be forced.) I don't think there is *nearly* enough emphasis, in many schools, on getting children to think actively about the mathematics as distinct from passively absorbing the diet, often attractively served and even pre-digested — which is fed to them. Maybe talk of conjectures conjures up the spectacle of children chaotically "discovering patterns", perhaps formulating wrong rules which may (horror!) be accepted or remembered by others. (Some readers may even be formulating unkind thoughts about impractical teacher-trainers.)

It is true that many of the conjectures offered by the children in Jeffrey's article were wrong, or partial; even so, for children of that age, they represented genuine and worthwhile attempts to come to terms with the problem. The teacher's task, as I see it, is to encourage children to probe, and ultimately to prove (justify), their conjectures.

Proof is the other side of the coin. Though I would want to encourage much more pattern spotting, I hope it doesn't end there, for the activity of mathematics involves testing, refuting, rejecting or establishing conjectures. A child's "hunch" is a starting point, but he has to examine it, check it, explain it, convince himself and others of its truth.

The ideas of proof have never been found easy — the illogicality of many arguments in everyday life (to say nothing of the misleading statements of advertisers and politicians) are witness to the difficulties and confusions which many adults find in dealing with such concepts as converse, implication, counter-example, etc. With the disappearance from the curriculum of formal Euclidean geometry, even children who do well in mathematics may have little contact with a way of thinking which characterises mathematics. This is not the place to develop these ideas, on which I have written elsewhere,[4] but I would argue that the particular quality which distinguishes mathematics from other forms of knowledge contributing to the curriculum is its particular emphasis on logical consequence. The characteristic statement of mathematics is "if A, then B".

So, *pace* IMA, let us not be deterred from trying to give children a wider view of mathematics than the arid "basic core", let us resist the public pressures from the possibly well-intentioned but generally ill-informed, and let us lose no opportunity to remind all who will listen that $M \supset N$: (Mathematics includes more than numeracy, whatever *that* may be!)

References

1. Fletcher, T. J. The continuing debate, *Mathematical Education for Teaching*, **3**, 1, July 1977.
2. Jeffrey, R. Making and testing conjectures, *Mathematics Teaching*, **82**, March 1978.
3. Simple sums floor vast majority of pupils, *The Times*, 9 March 1978.
4. Baxandall, P. R. *et al.*, *Proof in mathematics*. An occasional publication of the Institute of Education, University of Keele (75p post free). (A collection of material, intended for teachers, sixth formers, and undergraduates, illustrating the nature and variety of the idea of proof in mathematics, 130 pp.)

EDITORIAL

We are delighted to include in this issue an interview with the current President of the Mathematical Association, Professor Geoffrey Matthews. It becomes apparent that, during his career, he has been involved in a very wide range of activities. We wish him well with his latest venture into the mathematical education of children of nursery school age.

One of the encouraging things that Geoffrey Matthews sees coming from the Great Debate is the formation of the Assessment of Performance Unit (APU) which will begin monitoring national levels of performance in mathematics during 1978. In fact, the APU has a much wider brief than the testing of achievement in mathematics. Its terms of reference, laid down in the 1977 Green Paper *Education in Schools*, are "to promote the development of methods of assessing and monitoring the achievement of children at school, and to seek to identify the incidence of under-achievement". Initially, aspects of the curriculum to be monitored are mathematics, language and science. Work on personal and social development, aesthetic development and physical development is at an exploratory stage. Members of the unit are anxious to reassure us that the tests are neither intended to diagnose individual difficulties, nor will they identify schools or LEAs. Someone at the DES used the expression "benign tests".

Assessment of performance in mathematics within the APU is founded on *Tests of Attainment in Mathematics in Schools* developed by the National Foundation of Educational Research. (Here we have a further link with Geoffrey Matthews, as he reviewed a report of this work in *Mathematics in School*, Vol. 5, No. 5.) In 1978, tests of mathematics will be administered to pupils aged 11 and 15 years. It is expected that about 700 primary/middle schools and 500 secondary schools will be included, and that in each school about one-third of the age group concerned will be tested. The tests will reflect the use of mathematics across the curriculum, and the range of mathematical activities in the classroom. Part of the assessment will be a practical test. Reports should be available about one year after the surveys. A pamphlet giving further details of assessment in mathematics had recently been issued.

We pass no judgment until we see how the first tests have been conducted, and read the reports for the conclusions drawn. However, as teachers we must hope that the work of the APU will enable us to better understand pupils' needs, and to provide feedback on the effectiveness of our teaching in enabling children to learn and understand mathematics.

Volume 7 Number 1
January 1978

Index of Authors and Articles

[Note: Volume 1 appeared in 1971/2, Volume 2 in 1973 and thereafter one volume each year up to Volume 17 in 1988]
[The number in brackets indicates chapter in which article appears]